# The Ecosophical Paradigm

*Revisioning Life, Mind and Values
In the 21$^{st}$ Century*

# *The Ecosophical Paradigm*

### *Revisioning Life, Mind and Values*
### *In the 21$^{st}$ Century*

Bernie T. Scala

*To the universe, our home and parent*

*Human history becomes more and more a race*
*between education and catastrophe.*
  *~ H. G. Wells*


*Upon this gifted age, in its dark hour,*
*Rains from the sky a meteoric shower*
*Of facts ... they lie unquestioned, uncombined.*

*Wisdom enough to leech us of our ill*
*Is daily spun, but there exists no loom*
*to weave it into fabric.*
    *~ from "Sonnet," <u>Huntsman, What Quarry?</u>*
    *Edna St. Vincent Millay*

# TABLE OF CONTENTS

Introduction                                                    pg. 1


PART I          A NEW SCIENCE

   Anatomy of a Crisis/Hope of Deliverance          pg. 9
   The Roots of the New World View                  pg. 15
   The New Science                                  pg. 21


PART II         A NEW VIEW OF LIFE

   Nature is One                                    pg. 27
   Holons and Holarchies                            pg. 31
   Vitalism                                         pg. 35
   Mechanism                                        pg. 38
   Morphogenesis Explained?                         pg. 42
   Towards a Coherent Theory of Life                pg. 45
   Dissipative Structures and Hypercycles           pg. 50
   The Systems Theory of Life                       pg. 57


PART III        A NEW VIEW OF MIND

   The Santiago Theory of Cognition                 pg. 61
   The Gaia Hypothesis                              pg. 65
   An Intelligent Universe?                         pg. 72
   The Symbiotic Universe                           pg. 80

PART IV  A NEW SOURCE OF VALUES

*E Duobus Unum*  pg. 85
The Transpersonal Mind  pg. 90
Science and Values  pg. 96


PART V  THE NEW PARADIGM

Today's Crisis: Pathosis of a Misperception  pg. 103
The Ecosophical Paradigm  pg. 106
The Characteristics of the New Paradigm  pg. 110
The Educator's Most Important Role  pg. 114


PART VI  THE KOSMOS STORY

The Consolation of Meaning  pg. 119
In the Beginning  pg. 121
The Emergence of Life  pg. 126
The Evolution of Animals and Plants  pg. 130
Primates and Humans  pg. 134
The Development of Language  pg. 139


PART VII  THE MYTHICAL DIMENSION OF LIFE

The Universe as Our Home and Parent  pg. 147
*Tat Tvam Asi!*  pg. 152
Conclusion  pg. 157

Endnotes  pg. 163
Bibliography  pg. 199
Index  pg. 203

## *Introduction*

In 1992, over 1700 of the world's leading scientists, including the majority of Nobel laureates in the physical and biological sciences, signed the *World Scientists' Warning to Humanity.*[1] The declaration began as follows:

> Human beings and the natural world are on a collision course. Human activities inflict harsh and often irreversible damage on the environment and on critical resources. If not checked, many of our current practices put at serious risk the future that we wish for human society and the plant and animal kingdoms, and may so alter the living world that it will be unable to sustain life in the manner that we know. Fundamental changes are urgent if we are to avoid the collision our present course will bring about.

It then went on to list six areas in the environment (the atmosphere, water resources, oceans, soil, forests, living species) that they said were suffering "critical stress". They concluded:

> A new ethic is required — a new attitude towards discharging our responsibility for caring for ourselves and for the earth. We

must recognize the earth's limited capacity to provide for us. We must recognize its fragility. We must no longer allow it to be ravaged. This ethic must motivate a great movement, convincing reluctant leaders and reluctant governments and reluctant peoples themselves to effect needed changes.

Two decades later, this warning remains largely unheeded. Lawmakers are still indisposed to enact laws to avert the looming disaster, the majority of the people on this planet still do not understand the seriousness of the situation we are in or the dire consequences of inappropriate and timely action.[2] A point of no return is not far off if not already here, and ignoring this fact only makes the day of reckoning that much more certain.

If we are to wake from this sleep of reason, we must adopt a new ethic with respect to our relationship with and obligation to the rest of the biosphere. But every ethic needs to be structured in a world view. The question arises as to what world-view the new ethic can be founded upon.

In the past, religion served this role. It ordered the universe and placed humanity within that order. With the advent of the scientific revolution in the 17th century, the religious frame of reference was slowly but inexorably displaced by a secular-mechanist view of nature that translated to matters of belief and value, at least in the western world. Nationalism and communism, two side-products of that doctrine, had roles in spawning their own brands of ethic, but they proved to be more baleful than exemplary. The recent revival of religious fundamentalism is more a sign of dissatisfaction with the current value-system than a sound basis for a new one.

And so here we are in the latter days of a postmodernist world, self-absorbed and morally confused, where the primary concerns seem to be mindless amusement, the pursuit of profit, and ever-increasing levels of consumption. The result: in 2004 the International Union for Conservation of Nature (IUCN) estimated that species are becoming extinct at a rate one thousand to ten thousand times greater than it would be without humans.[3] The 2005 UN Millennium Ecosystem Assessment affirmed that human activity had seriously degraded 60% of the planet's ecosystems.[4] The 2010 Economics of Ecosystems and Biodiversity Report declared that 85% of the oceans and seas are adversely affected.[5] We are in the midst of what has been called the Sixth Great Extinction,[6] this one unique in that it is caused by the activities of a single species: us.[7]

We, moreover, are changing the planet's climate at a faster rate than has ever occurred.[8] This is already leading to unprecedented melting of the permafrost and ice caps,[9] changes in the flow of ocean currents, the disruption of weather patterns, extreme temperature fluctuations, increasingly frequent and fierce wildfires and freak storms.[10]

And this is only the prelude of much worse to come, according to even moderate computer simulations extrapolating from present trends. We cannot stop these processes — they are now ingrained in the natural dynamics that operate on this planet. The best we can do is slow them down. But our current policies and actions, if anything, are causing them to gather momentum. Recent news reports tell of a 27% jump from August 2010 to April 2011 in Brazil's deforestation rate.[11] And the International Energy Agency has just reported record worldwide carbon-dioxide emissions in 2010.[12]

If we are to step back from the brink, or at least mitigate the outcomes predicted, it is imperative that we learn to see ourselves as an inextricable part of a whole with which we share a common destiny. This new perspective, which I term the Ecosophical Paradigm, is a radical one. But were we to adopt it, it could sufficiently transform humanity as to steer us a little away from a future that threatens to be much impoverished and far less pleasant than we have been accustomed to for millennia.

This change of consciousness will come not in a mystical flash, nor from the lips of some charismatic figure. Neither would be considered adequate grounds on which to found a new ethic, anyway. Rather, the change of consciousness will need to be acquired purposely, systematically, and will need to be legitimized by the methods and principles of science. These alone would be deemed sufficiently rigorous by the standards of the 21st century mind for grounding a universally acceptable new *Weltanschauung*.

At the same time, humans are not cold calculating machines that feed on facts and always act rationally. We have interior lives subject as much to passion and personal quirks as to logic. For us to be won over we need to be reached emotionally, and the best way to do this is by means of art. It is through the aesthetic experience that humans most effectively internalize complex ideas, render them immediate, meaningful, and cause for action. The wisdom that could leech us of our ills is slowly being spun into a tapestry of poetry and power beyond anything imagined by the greatest bards of fantasy. Yet it is real. To be open to its message is to see with new eyes.

This book is an introduction to a new world view long intimated but too slow to materialize. It sketches the intellectual history that led to the emerging world view, its key ideas, and the means by which it can be popularized. It addresses the missing dimension in modern education, *meaning*, and it proffers a course by which a sense of proportion and sacredness can be put back into our lives.

It is also a plea, borne of grave concern and exasperated impatience, to people of goodwill everywhere to stop and seriously consider the situation we are in. How much more evidence do we need? When will we wake up to the clear and present danger before us? The facts are indisputable, regardless what the few self-serving naysayers who earn their living by being controversial say.[13] We can no longer ignore the omens. Nor can we continue to be so willfully irresponsible as to be the destructive agents of so much in this world of wonders, and perhaps of ourselves in the process. This is neither scaremongering nor alarmist nonsense. Check the facts, weigh the evidence.

The economic meltdown of 2008 shows how quickly matters can get out of hand. It should be a lesson on the importance of preventing disaster before it strikes. Time is running out and the calls for decisive action are becoming evermore numerous, evermore insistent. If we are to deal with the formidable problems before us, we must come together, reason together and act before it is too late. We have to learn to set aside our differences while we grapple with the danger that threatens to bury us all. It is only by mobilizing as a species that we stand any hope of success. There are no more excuses. We *must* confront the issue and commit ourselves to the challenge it poses with all the earnestness and genius that we are presently directing towards armaments, production and consumption.

The means and tactics we will employ in this endeavour I will leave to those more qualified to prescribe than I am.[14] The important point is that we act. The *status quo* is no longer an option.

How we respond in this moment of supreme danger will define us as a species worthy of either cosmic respect or shameful scorn for having squandered so spectacularly our considerable endowments. I trust that our better sense will prevail and guide us towards wise and appropriate action. I hope you join me in spreading the message — there is no more urgent task at hand.

# PART I

## A NEW SCIENCE

### *Anatomy of a Crisis / Hope of Deliverance*

Our world view[i] is changing. For four centuries the Cartesian-Newtonian vision of a mechanistic, material reality has dominated the scientific world view which, in turn, has permeated all aspects of western culture. Now a shift is occurring, perhaps one of the most fundamental and dramatic transitions in history. It is nothing less than a basic change in the cultural belief structures of western industrial society. A global mind change is upon us. (Michael Thoms, *New Dimensions Radio*, 1988)

The 20th century was a time of unprecedented ferment in the intellectual history of humanity: psychoanalysis, the Theory of Relativity, Quantum Mechanics, Gödel's theorems, cybernetics, Information Theory, the deciphering of the genetic code, Bell's Theorem,

---

[i] A world view is a conceptual system by which an individual, society, or culture makes sense of the world. It can be thought of as metaphors by which we live. Our world view not only structures our perceptions, but gives us a frame of reference by which we can orient ourselves in our world, guide our actions, and give meaning to our lives. (see Sinetar: 44)

Complexity Theory, Systems Theory, Chaos Theory, Artificial Intelligence, virtual reality. Each of these contributed to a substantially richer understanding of the cosmos and our place in it.

As well, there arose, especially in the second half of the century, the growing realization that our dominion over the Earth must be reappraised and become more responsible. The new world view referred to by Michael Thoms above is a result of the convergence of the century's many scientific insights and the newly-acquired ecological sensibility. It is radically different from the Cartesian-Newtonian view which has had hegemonic dominance for over four hundred years. It is important to know how this new perspective has come about and why we must expedite its adoption.

From the time of Francis Bacon (1561-1626), the western view of nature has had an anthropocentric (indeed, *andro*centric)[15] orientation in which progress entailed mastery over it and exploitation to the fullest of its resources. It is now clear, however, that the assumption that the world was made only for our own advantage must be reconsidered. Climate change, vast deforestation, mass reduction of biodiversity, the alarming rate of soil erosion, the deterioration of the ozone layer, the melting of the ice caps, the depletion of aquifers, the spread of coastal "dead zones", etc. in addition to an ever-increasing world population (that, although slowing in rate of growth, will have gained another billion in 20 years),[16] plus a widening gap in the distribution of wealth and opportunity (developed nations, with barely 20% of the world's people, control 86% of its wealth, consume 80% of its resources and create 83% of its waste)[17] have led to the conclusion that unless we change our ways very soon, we will face cataclysmic environmental and socio-political consequences.

According to the "ecological footprint" model for measuring the allocation of resources, humanity is presently using the equivalent of approximately 1.5 planets in terms of the natural assets we expend and for absorbing the waste we produce.[18] This means that it now takes the Earth one and a half years to regenerate what we consume in one year. Moderate UN scenarios suggest that if current population and consumption trends continue, by the 2030s, we will need the equivalent of two Earths to support us. And if everyone were to have the same standard of living as that enjoyed in the first world today, we would need *four* more planets like ours.[19] "[T]o believe that things are still well in the world one must ignore three-fourths of it," has said Robert Kaplan of the *Atlantic Monthly*.[20]

Maurice Strong, former under-secretary general of the United Nations and one of the most respected public servants of our time, stated at the turn of the millennium that we had about 20 years in which to act before events begin to spin out of control.

> It is not that the demise of our civilization could occur rapidly but that the decisions and actions that would determine its ultimate fate are likely to emerge within the first part of the new century, and particularly within the next two decades. (Strong: 359)

The general consensus among the scientific community is in accordance with this alarming prognosis. The tragedy is that it is already too late for a good deal of the biotic community with which we share our planet.

> [We are] bringing about the "greatest impasse to the abundance and diversity of life on Earth" since the earliest beginnings of life some four billion years ago. In the estimate of Paul Ehrlich, eminent biologist, we are probably extinguishing some ten thousand species each year. (Berry and Swimme: 247)

The average extinction rate is now some
1,000 to 10,000 times faster than the rate
that prevailed over the past 60 million years
… Only a small fraction of the world's plant
species has been studied in detail, but as
many as half are threatened with extinction
… The greatest threat to the world's living
creatures is the degradation and destruction
of habitat, affecting 9 out of 10 threatened
species.[21]

Ecosystems destroyed, wild climatic fluctuations, countless species forever lost, millions of our own dead or displaced, wars, social turmoil, financial losses unlike anything ever witnessed, are all part of the scenario foreseen. Already the scramble for diminishing resources and related rivalries is leading to the hardening of geopolitical stances.

"The era of good feelings associated with
the heyday of globalization has gone forever.
Something grimmer, less productive, and
less predictable has taken its place."[22]

The sobering conclusion is that the heretofore presumption that the environment is a commodity for the appropriation of whoever is first able to exploit it, has to be replaced with one that can accommodate human interests within the context of the needs of the larger biotic community. We have to shift our persuasion from one of profligate consumption to one of consideration and sustainability. We need, in short, a global ethic respectful of nature that would bring about the kind of social changes needed to pull us out of an ever-worsening spiral. The alternative is disaster.

> Current world conditions call for a unified global approach with value perspectives built on something higher than just the human species or its societal dynamics, something more godlike that will include the welfare of the total biosphere and ecosystem as a whole on an evolutionary time scale. The greater the human impact on the ecosystem, the more urgently these higher perspectives are needed. They are imperative also in efforts to perceive higher meaning, where it becomes a logical necessity that humanity be able to perceive itself in terms of a meaningful relation to something more important than itself. (Sperry: 1985, 73-74)

For the receding Cartesian-Newtonian world-view, a key idea was unlimited material progress achievable through continuous economic and technological growth. In the emerging world view, this assumption has been replaced by the notion of *sustainability*, which Lester Brown, founder of the Worldwatch Institute has characterized in terms of satisfying a society's needs without diminishing the prospects of future generations.

## *The Roots of the New World View*

As mentioned, ecological sensibility is one of two roots of the new world view. Its genesis can be traced to the 1962 publication of Rachel Carson's *Silent Spring*, an impassioned and detailed report on the damaging impact of chemicals, especially D.D.T., on the environment. But it took hold of public attention with the "Earthrise" photos taken during the Apollo missions. It was the first time in history that we could see our planet from a distance, a blue-white gem floating in a void of limitless space. In these photos we learned to recognize the beauty and fragility of our home. They proved to be a powerful symbol for the burgeoning "green movement," and may well have been the most significant result of the entire space program. (Capra: 1982, 284)

Concerns about pollution, "the population explosion," the greenhouse effect, nuclear winter, the wide-scale destruction of rainforests, etc., plus the rapid expansion of telecommunication technology that made neighbours of people from opposite poles of the globe, all contributed to an increased concern for our planet. The 1972 report *The Limits to Growth* by the Club of Rome, a global think tank that deals with international issues, became a best-seller. The ecological movement was well on its way to becoming a social and political force.

The suddenly-recognized need to counter the negative impact of humankind on the environment led to the creation of "green" political parties. The first formed in Australia in 1972 to fight deforestation and dam building in Tasmania. Within a few years, dozens of other Green parties all over the world had candidates running for elections. Although their political success has been mixed (the German Greens a notable exception) environmental initiatives have focused and have had most influence on the local level ("think globally, act locally"). Most of their tenets, furthermore, including respect for diversity, sustainability, participatory politics, community-based economies, and social justice have now been co-opted by parties across the political spectrum. Interest in the environment has become mainstream, but not to a sufficiently effectual degree, as evidenced by our current problems.

The other root of the new world view is an essentially different scientific understanding of reality and mind than the Cartesian-Newtonian world view which, although critically important for replacing the Aristotelian-Scholastic accounts of nature with empirical science, had become, by the late 1800's, inadequate for explaining new experimental results.

In his 1905 Special Theory of Relativity, Einstein showed that when it comes to describing the motion of bodies in space, the observer is an indispensable component of that description. This was in stark contrast to classical mechanics, which had assumed that the physical world could be accounted for in completely objective terms, as though there were some absolute standard for measuring time and space. In Relativity Theory, no such standard can exist. Every observer operates from a particular *frame of reference* from which he or she bases measurements, and this subjective perspective[23] constitutes an essential feature of all descriptions of any physical phenomenon. The upshot was that this new way of considering reality was able to explain phenomena that the Cartesian-Newtonian view could not.

While Einstein was redefining the meaning of space and time,[24] other physicists were being forced to re-examine their understanding of the behaviour of atomic particles. John Dalton's atomic theory was successful in explaining most of the experimental results pertaining to the micro-world, but it could not account for the observation that atoms do not emit heat or light energy at a consistent rate for all temperatures. In order to come to grips with this puzzling fact, Max Planck in 1900 was compelled to posit the notion of the *quantum*.

Planck asserted that objects absorb or emit energy not gradually and smoothly, as was assumed in classical physics, but in discrete bursts known as quanta. Consequently, it is impossible to know, let alone predict with any degree of accuracy, the behaviour, momentum or location of any particular subatomic particle,[25] say an electron. Only the mass behaviour of large numbers of such particles could be accurately accounted for.

The laws that were developed from this insight are markedly different from those of Cartesian-Newtonian mechanics, which had held that one could, in principle, determine perfectly the behaviour of any component in a system given sufficient information of the initial states of every constituent of that system. The new laws no longer provide certainty or definitive answers; all they can offer is probabilities, approximations, statistical descriptions.

> Scientific theories can never provide a complete and definitive description of reality. They will always be approximations to the true nature of things. To put it bluntly, scientists do not deal with truth; they deal with limited and approximate descriptions of reality. (Capra: 1982, 48)

The loss of certainty became even more pronounced when it was discovered that in studying matter at the subatomic level, the very act of observation changes the results of the experiment.[26] Thenceforth, our understanding of nature was seen to be largely subject to the method of investigation we use.

> Within this context, consciousness does not just passively reflect the objective world; it plays an active role in *creating* reality itself. (Grof: 6, emphasis mine)

For example, at its most fundamental, the physical world can be either continuous (wave-like) or discrete (particle-like) depending on what apparatus we choose to examine it with. Again, as Einstein had shown with the macro-world, so it is with the micro-world: the observer is an essential feature of the phenomena observed because there is no sharp boundary separating the two. Reality is no longer a given "out there" but a construct of the observer interacting with the observed. When we measure something, we are forcing an undefined world to assume a definite value, and in so doing, "measurables" are replaced with "observables". In short, we cannot know nature, we can only *experience* it. There is no theory-independent reality that we can know, and what we know is largely constructed by us.

> The science of nature does not deal with nature itself but in fact with the *science* of nature as man thinks and describes it. (Heisenberg, as quoted in Haberman: 141)

A physical law, then, is not a comprehensive account of some regularity that exists in the world (as in Newton's clockwork universe), but a descriptive model, generally in mathematical terms, that imposes a conceptual framework upon phenomena observed.[27]

### The New Science

Ironically, while physics from the 1920s onward moved further and further away from its materialist, determinist roots, other sciences, including the social sciences, strove to become more like classical physics in an effort to establish materialist, determinist, mechanist bases for their disciplines. An example of this approach in psychology was Behaviourism, whose fixation with objective explanations led to the denying altogether of consciousness from its account of behaviour. As misguided as that and similar approaches now seem, the dogma of positivism had such tight grip on the scientific establishment that few seemed to question it.

All this changed with the 1972 publication of *Objective Knowledge: An Evolutionary Approach*, in which the Austro-British philosopher Karl Popper called for a new way to do science. The kind of science that works well at explaining the physical world, he argued, does a poor job at explaining the biological or human worlds. The reason is that the physical sciences (physics, chemistry, earth sciences) are essentially materialist, reductionist and positivist, whereas the biological and social sciences are organic, holistic and qualitative.

Although the differences in kind between the physical and the lived worlds had been recognized as early as Giambattista Vico (1668-1744) and vigorously celebrated in the Romantic reaction to the Enlightenment, it was Popper's reputation as a tough-minded philosopher of science that caused present-day intellectuals to take notice. The entire scientific enterprise was suddenly open to whole-scale reappraisal.

Whereas in the past, science would stake out a territory and disregard anything that did not fit its strict criteria (measurement, strict experimental methodology, formalized theories, etc.) for what it deemed valid areas of inquiry, now *any* phenomenon could to be considered a proper starting point for science. Its new role was to develop a rigorous and consistent way of dealing with any phenomenon on its own terms without prejudice.

> This adequate epistemology will be, above all else, humble. It will recognize that science deals with *models and metaphors*

> *representing certain aspects of experienced
> reality*, and that any model or metaphor may
> be permissible if it is useful in helping to
> order knowledge, even though it may seem
> to conflict with another model that is also
> useful. (The classical example is the history
> of wave and particle models in physics.)[28]
> (Harman and Sahtouris: 26)

Thus, whereas before the only valid science had been reductionist (i.e. intent on simplifying to, essentially, materialist terms), objectivist (assumptive of an external world that could be studied free of any subjective bias), and positivist (dealing only with what could be measured and, ideally, mathematically formalized), science could now be unhindered in its choice of subject matter or rubric of inquiry. In short, no area of human experience, including the paranormal and non-ordinary states of consciousness, need henceforth be excluded from scientific study.

The practice of science, furthermore, need no longer be empirical (i.e. based on the results of strictly parameterized experiments), but analogical (founded on correlations and resemblances among diverse phenomena). Its modes of explication need no longer be analytic (intent on explaining a phenomenon by breaking it down to its most irreducible components) but structural (appreciative of phenomena in their wholeness). And its approach need no longer be foundational but "bootstrap," that is, assumptive that no concept or law is more fundamental than, or even logically prior to, any other because the meaning of any part of a theory depends on the theory in its wholeness.

This new *modus operandi* does not imply that science would be any less disciplined or less scrupulous in its search for explanation; merely, that it could now relax its self-imposed strictures because it is now accepted that different areas of study require different methods. The model of physics as the science *par excellence* is no longer pertinent to the enterprise of scientific investigation. The ideal now is open-mindedness, dialogue, a multidisciplinary approach to problems, and mutual respect among the various fields because it is understood that objectivity is a social construct obtained through multiple perspectives. Each discipline must find its own ways and means to deal with its areas of inquiry, while at the same time acknowledging that other disciplines could help illuminate the subject matter in its field.

Amidst this new scientific *Zeitgeist*, Nobel laureate Roger Sperry published (in 1987) *The Structure and*

*Significance of the Consciousness Revolution in Science.* It caused a stir. Working as a neurophysiologist studying the so-called "split-brain" phenomenon, Sperry, contrary to his orthodox materialist leanings, had become convinced that, not only is consciousness[ii] metaphysically real, but that the generally accepted explanation of it as an epiphenomenon reducible to matter (the brain) was wrong.

He also realized that there are two kinds of causation in the world. There is "upward causation" whereby atomic and subatomic particles jostling randomly about produce effects in the macroscopic world. (For example, hydrogen and oxygen atoms combine to form water molecules, which combine to form clouds, fall as rain, etc.). But there is also "downward causation" of the kind that mind can intentionally bring about. (For example, human beings can seed the atmosphere with carbon dioxide pellets in order to make it rain so as to nourish plants for animal feed.)

---

[ii] Although stubbornly eluding a scientifically rigorous definition, "consciousness" is generally understood to have something to do with attention, awareness, self-reference, interior dialogue, the ability to feel or experience, a sense of physical and/or mental presence, the intuition of selfhood, and the like. John Eccles' concept of the *psychon*, the theoretical elementary unit (quantum) of consciousness, is deemed by some as a possible interface between physics and psychology.

Science since the Enlightenment had recognized only the first kind of causation and had explained away the second kind as reducible to the first, that is, to the firing of neuronal synapses in the brain brought about by imbalances in the electrical charges of certain ionized molecules (neurotransmitters). For Sperry, the *intentional* aspect of consciousness and its role as the unifying factor behind diverse mental states, cannot be adequately accounted for in this way. Nor can the *consequences* of the operations of a conscious mind, as is evident in the present ecological crisis. In fact, for Sperry, the human mind is now the single most powerful causal agent in the world. "The human brain is today the dominant control force of our planet." (Sperry: 1985, 10)

> No other causal system with which science now concerns itself — earthquakes chemical reactors, magnetic fields, you name it — is of more critical importance [than the human brain] in determining our future. (Sperry: 1983, 71)

What Sperry considers the logical conclusion of this insight will be dealt with in a later chapter.

# PART II

# A NEW VIEW OF LIFE

## *Nature is One*

Popper's call for a realignment of scientific inquiry, and Sperry's emphasis on the causal efficacy of consciousness over matter led to the "cognitive" or "humanist" revolution in science in the last decades of the 20th century. Essentially, what the New Science offers is a wholly new perspective of nature and mind. One of its principal assumptions is the interrelatedness, indeed *interdependence* of all experienced phenomena. This radical unity is now recognized as being grounded in the very fabric of the universe. Here is how this startling conclusion was arrived at.

In the early 1980s, the French physicist Alain Aspect provided strong[29] experimental evidence that a quantum event in one locus can *instantaneously* affect an event in another locus without any feasible means of communication between the two loci. (Einstein had doubted the possibility of this effect, calling it "ghostly action at a distance"). Since, according to the Theory of Relativity, no physical object can travel faster than light, the fact that events could influence each other before any information could be transmitted between them appeared to confirm one of the two mutually exclusive inferences of Bell's Theorem:[iii] the Cartesian-Newtonian conception of reality as consisting of separate objects bound to local conditions is false.

What the result of Aspect's experiment implies is that every individual particle in the universe is interconnected with every other particle in a single field of communal and immediate reciprocity such that any change in any one location is immediately felt throughout the universe. In other words, not a leaf falls from a tree without it instantaneously affecting even the furthest galaxies and everything in between. Thus, rather than there existing discrete objects and empty spaces between them, as common sense and science have held for centuries, it is now understood that everything is part of a universal substratum

---

[iii] In a nutshell, Bell's Inequality Theorem states that if we are to have a scientifically rigorous explanation of the universe, then either the assumption of *reality* (i.e. that material objects have an objective existence totally independent of the observer), or that of *locality* (i.e. that a phenomenon in a specific space-time location cannot have any effect on an object in some sufficiently distant space-time location), must be false. Hence, our everyday notion of the world, which is grounded on both assumptions, is radically mistaken: Reality is not as we think it is.

of varying densities of matter-energy. In physicist David Bohm's words, the cosmos constitutes "an unbroken, coherent whole." The appearance of separate parts conceals an "implicate order" by which not only is everything in perpetually intimate communion, but "in some sense each region contains [the] total structure 'enfolded' within it."[30]

For Bohm, the universe can be thought of in terms of a *hologram*,[31] a three-dimensional photographic image produced by the interference pattern of two slightly off-phase laser beams (i.e. light beams of a specific wavelength/frequency). The essential feature of a hologram is that the holographic image is diffuse throughout the photograph and thus, any fragment of the hologram will contain a smaller but otherwise complete image of the entire photograph. The key point is that every part of the hologram, however divided, has access to the whole or, to put it differently, that the whole is "folded into" every part. So it is with the universe: the universe is an infinitely ramifying/exfoliating process of inexhaustible possibilities whereby an "implicate realm" of undivided wholeness is embedded in the "explicate realm" of separate things and events of experience.

What at the subatomic level seem to be randomly-occurring events (particles popping in and out of existence, colliding to produce other particles, disappearing here to reappear elsewhere, etc.) are in fact determined by their non-local connections to everything else in the universe. Consequently, whereas in classical mechanics the properties and behaviour of the parts determine the behaviour of the whole, contemporary physics tells us that the whole also determines the behaviour of each and every constituent part. Paralleling Sperry's characterization of mind, events in the physical world are affected both from above (more complex and encompassing levels) and from below (less complex and contained levels). This new ontology is categorically monistic.

> The universe is no longer seen as a machine made up of a multitude of separate objects, but appears as a harmonious indivisible whole; a network of dynamic relationships that include the human observer and his or her consciousness in an essential way. (Capra: 1982, 47)

## *Holons and Holarchies*

The universe as a multi-layered, intrinsically dynamic web of relations concurs with how contemporary biology views the living world. The Cartesian-Newtonian model of nature as clockwork, with non-human organisms, including primates and the family pet, as machine-like automata determined by the same laws as are billiard balls, has now been replaced by a holistic or systems theory of life.

Nature is presently understood as a growing, evolving, endlessly creative process of perpetually interconnected ecosystems. While all organisms are wholes in the sense of being integrated structures, they are simultaneously also parts of larger, more complex wholes which, in turn, are parts of still larger wholes, and so on, all the way up to the cosmos in its entirety.[32] The Hungarian-British essayist Arthur Koestler coined the word "holon" to refer to these whole/parts.

All natural hierarchies, then, are *holarchies* consisting of both "depth" (vertical) and "span" (horizontal) dimensions, and in which there is an asymmetry of increasing unity and integration among its constituent holons. For example, protons contain quarks but quarks do not contain protons; atoms contain protons but protons do not contain atoms; molecules contain atoms but atoms do not contain molecules, etc.

Each successive holon incorporates the lower holons in its structure and integrates them into a new entity. While these modules act autonomously, the relationship they have among each other affects the complex and brings forth *emergent properties*[iv] that are qualitatively different than anything that came before. For example, the wetness of water is a feature neither of hydrogen nor oxygen nor, for that matter, any single water molecule, but a quality that emerges when a sufficiently large number of $H_2O$ molecules are in close proximity to each other under certain conditions of pressure and/or temperature.

Or, take an example from biology. Place a termite in a suitable environment and it does nothing. Add another termite and another still. Nothing changes. But keep doing this, and in due course, when a critical mass is reached, the termites spontaneously organize into a colony and set about

---

[iv] The term was used in the early 1920s by the philosopher C.D. Broad to refer to those properties that arise at a certain level of complexity but do not exist at lower levels.

the task of building a city for themselves. Like the wetness of water, the characteristics of the termite colony emerge as totally original properties of a new structure.

The fact that each holon is a whole/part assigns it two roles. In order to exist and preserve its particular autonomy, each holon must assert its identity. At the same time, because it is part of a complex, it must integrate with all other proximate holons, horizontal and vertical, so as to function as part of the larger whole. Hence, every holon must maintain not only its *agency*, but also its *communion* with the extensive net of relations upon which its existence depends. These two tendencies are opposite but complementary and in any functional system, whether it be an individual entity, social group, or ecosystem, there is a dynamic balance between self-assertion and integration. If any holon unduly compromises either its communion (as a part) or its agency (as a whole), it ceases to exist, and whatever remains of its structure is incorporated into other holons. If for example, the termite colony grows too large too quickly for its habitat to support it, it will run out of food and perish. Its components will themselves be absorbed as food by other organisms in the holarchic complex it was part of.

Thus, we see that while each holon plays its particular role as an independent unit, because it is ensconced in the structure of a larger whole, its autonomy is restricted by its *holonomy*, that is, the autonomy of the outer, larger holons. To survive, to be a viable entity, the entire holarchy must synergize into a mutually supportive complex. A holon's vertical capacity, then, is one of self-transcendence and self-dissolution simultaneously. It can cease to be and dissolve into something simpler, or it can combine with suitable partners to bring forth new structures and new qualities.

> In general, for any particular holon, *functions and purposes* come from levels farther out in the holarchy; *capabilities* depend upon the next level in. There is no unique holarchy for a given holon; for one purpose one may want to speak of organisms within species, and for another purpose, of organisms within communities. Some holons may not seem to have a function with respect to the next holon out — parasites in a larger organism, for instance — yet may serve a function in some holon still further out, such as the ecosystem. (Harman and Sahtouris: 18)

In the depth-span dimensions of the Nest of Being, there somewhere emerged the new quality called *life*. This new feature, according to C. Lloyd Morgan, one of the 20th century's earliest proponents of biological holism, cannot be predicted nor explained in terms of its elementary holons, organic molecules. Life is an *emergent property* of matter, neither reducible to it nor wholly distinct from it. A dualist[33] ontology, for Morgan, is not needed to explain life. Thus we see the germ of an alternate conception of life than that of Mechanism and Vitalism, the hitherto major theories of life.

### *Vitalism*

Vitalism, a doctrine that has roots in animist pre-scientific thought, was first given rational footing by Aristotle and his followers. For the Aristotelians, every living organism has a soul, which gives form to the body and determines its *telos* (goal or purpose). In its modern guise, Vitalism became prominent in the late 19th century in reaction to scientific materialism and Kantian idealism. It maintains that the patent difference between living and non-living bodies is due to a non-material "vital" principle that animates organisms and enables them to be in some measure self-determining.

While vitalists differ in the details, they share a number of common beliefs. These include: a dualist ontology; the irreducibility of biology to physics; the view that lived reality consists in movement and development rather than static being; that reality is essentially organic rather than mechanical; and that the functions of living organisms as integrated wholes cannot be understood from a study of their constituent parts alone.

A physical system, a machine, for example, is nothing more than the sum of its parts and the interactions between them. A living organism, by contrast, has a wholeness that so transcends its constituents and their interactions, that it can often regain normal form even if some of its parts are damaged or removed. There is something in every organism, in other words, that is inherently integrative and essentially teleological (end-oriented) in a way that no machine is.

This fact was made dramatically clear by the discovery of Hans Driesch, one of the most influential vitalists of the 20th century, that despite even extreme tampering in the early stages of an embryo's development, organisms still develop into complete, perfectly formed adults. He concluded that there has to exist some matter-transcending vital element that, like Aristotle's "forms", controls and promotes the growth and development (morphogenesis) of an organism. He chose Aristotle's term *entelechy*[34] to refer to this teleological principle. The entelechy of an organism, for Driesch, is a non-mechanical principle or agent (to Aristotle, the soul) that directs the psychochemical processes of life without being part of them, and contains within itself the blueprint of the adult organism that, in modern terminology, "attracts" the developing organism towards that form.

Although Vitalism appealed to many thinkers including Dilthey, Nietzsche, and Bergson, its dualistic assumption and its inability to provide an adequate account of *how* the entelechies are able to carry out their supposed roles kept it from being widely accepted. It was, rather, Mechanism that became the standard dogma among most biologists in the early part of the 20th century.

## *Mechanism*

Mechanism, another venerable theory of life has its roots in the Pythagorean axiom of the eternal truths of mathematics, and the atomistic tradition of materialism. It is based on the methodological thesis that genuine scientific explanation of nature involves mathematically precise mechanical models describing cause and effect in terms of basic material entities coming in direct contact with other material entities. The discovery of the biochemical basis of biological processes, the formulation of the laws of heredity, the explicative power of the germ theory of disease, etc. were all cited as supporting a mechanistic explanation of life.

Notwithstanding physics' splendid success with a mechanical model of the universe, the machine analogy breaks down when it comes to understanding the growth and development of organisms, their morphogenesis. Apple trees grow from apple seeds. Mice, humans, and whales develop from small fertilized eggs that are practically indistinguishable from one another or, for that matter, from the eggs of any other animal. As the zygote divides again and again, each new cell contains identical genetic material.

Yet some develop into liver cells, others become blood, still others limbs. (There are over 200 types of cells in the human body). No machine grows spontaneously out of seeds or eggs. Nor do they reproduce to bring forth new machines from small parts of themselves. They do not regenerate after damage. If a hydra or a flatworm, by contrast, is cut into pieces, each part can still develop into a complete organism. If dozens of cuttings are made from an elm tree, each can grow into a new tree. No machine can do that. Nor can a machine fix itself if damaged, whereas if an animal is wounded, its body immediately begins to undergo the process of healing. And then there is the astounding complexity of organisms: even the simplest life form is far more sophisticated than the most advanced products of any human technology. In these examples and countless more, it is clear that the machine analogy for living organisms is inadequate.[35]

In order to counter such criticisms, mechanists have resorted to a number of different explanations for the manifestly teleological disposition of life. In the late 19th century, the inner organizing principle of organisms was identified with the germ plasm inside cell nuclei. The nucleus was then understood as a tiny brain directing and controlling the body of the cell around it.

With the discovery of the structure of the DNA molecule in 1953 by Crick and Watson, genes, the carriers of genetic information that is passed from generation to generation in the sex cells of all organisms, were deemed by mechanists to be the vitalizing agent of all cells. Far from being just molecules of purines and pyrimidines, the genes were endowed with all of the properties of life and mind. The biologist Richard Dawkins[36] has gone so far as to suggest that organisms are "throwaway survival machines" built by "selfish genes" to live in and be conveyed through to future generations.

> These genes are no longer mere chemicals; they have come to life and have minds like ruthless men. Not only do they have powers to 'create forms,' 'mold matter,' and 'choose,' but they engage in 'evolutionary arms races' and even 'aspire to immortality.' (Sheldrake: 1992, 100-101)

It became clear over time, however, that genes merely provide instructions for the cell's organelles to produce particular proteins for particular cellular processes.

They cannot account for the seemingly purposive characteristics of cells and particularly their morphogenesis. In other words, genes code for the various materials that make up an organism (analogous to the brickwork, plumbing and wiring of a building) but not its overall plan (analogous to the building's blueprint).

With the advent of information technology, a new mechanistic model was proposed in terms of computer software. The teleological, organizing principles of organisms were considered "genetic programs". But, as critics pointed out, computer programs originate in human minds and are purposive because they are consciously designed for some end. The notion of a genetic "program" organizing and directing organisms suggests that it is intentional and mind-like, but who is the author of these programs? How can genes, which are no more than molecular rungs in the DNA double helix, be supposed purposive, self-interested, "ruthless"? Appealing to concepts like "genetic programs" or "selfish genes" provides no better explanation than attributing life processes to entelechies. In this light, Mechanism appears less an alternative theory than a cryptic form of Vitalism.

## *Morphogenesis Explained?*

Not content with pointing out the deficiencies in the mechanist account, the contemporary neovitalist Rupert Sheldrake has proposed the Hypothesis of Formative Causation. He argues that no matter how sophisticated a materialist theory may appear, it cannot explain the creative forces behind the order and transformations we see everywhere in living nature. To him, embryological development (as well as instinct, behaviour and memory) can best be explained in terms of "formative causation," the basis of which are *morphic fields*, "a kind of collective memory[37] on which each member of the species draws, and to which it contributes." (Sheldrake: 1992, 110)

He posits that autonomous systems at all levels of complexity (in other words, holons) are able to organize themselves into certain standard patterns because their morphic fields (analogous to the fields that surround a magnet) are influenced, through "morphic resonance," by the morphic fields of other systems resembling them. The stronger the resemblance, the more pronounced this influence, which, in turn, becomes amplified by the resemblance, the entire system locked in a virtuous feedback loop. Each species has its own form-field and each organism is the intersecting nexus of multiple fields. The human body, for example, has a morphic field as a whole. But there are also fields for each limb, organ, cell, etc., as well as for our behaviours and beliefs. Each of these fields renders a species and individual organisms both unique and universal. The regularities we see in organisms are due to certain dispositions or "habits" that are conveyed across space and time by means of morphic resonance suitably attuned to specific morphic fields. In short, similar things influence each other over space and time.

Organisms, then, inherit not only genes but also fields of organization by means of which their morphogenetic development, their pattern of behaviour, the rapidity with which they learn or adapt to new situations, etc. are drawn towards certain goals or ends. These "attractors," have been established by remote ancestors and reinforced by each and every individual in some harmony with the morphic fields. When something new is experienced by an individual, other members of that and related species find it easier to respond appropriately to it under similar situations.[38] With respect to the morphogenetic development of organisms, genes define the range of parameters within which these developments take place, but the principles of the organizing processes themselves are particular to the morphic fields that a given entity both resonates with and amplifies.

> I am suggesting that heredity depends not only on DNA, which enables organisms to build the right chemical building blocks — the proteins — but also on morphic resonance. Heredity thus has two aspects: one a genetic heredity, which accounts for the inheritance of proteins through DNA's control of protein synthesis; the second a form of heredity based on morphic fields and morphic resonance, which is nongenetic and which is inherited directly from past members of the species. This latter form of heredity deals with the organization of form and behavior.[39]

Sheldrake has provided a number of fascinating examples from biology and other fields to support his position.[40] His hypothesis would also account for such baffling phenomena as the seasonal migrations of monarch butterflies,[v] the complicated life cycle of parasites, the

---

[v] The North American variety travel thousands of kilometers, some from far away as Canada, to congregate in certain mountainous regions of Mexico without ever having been there before.

ingenious subterfuge of carnivorous plants, the nest-building behaviour of birds and insects, asynchronous germination,[41] etc.

Nevertheless Sheldrake's notion has generated more debate than gained supporters. Most scientists and philosophers are wary of Vitalism in any form because of its dualistic presupposition. As well, the existence of morphic fields appears difficult if not impossible to verify scientifically — we have no inkling of any instrument that could possibly detect them. Nor are they concepts like quarks which, although not measurable, fit an experimentally confirmed, well-supported theoretical scheme that can generate testable scenarios. Hence, for many thinkers, including physicist/systems theorist Fritjof Capra, the contemporary Systems Theory of Life offers the best alternative to both the mechanist and vitalist accounts.

### *Towards a Coherent Theory of Life*

Systems Theory agrees with Vitalism's position that the phenomenon of life is central to experienced reality and much too rich to be adequately accounted for in mechanistic terms. It, in turn, agrees with Mechanism's position that an explanation of life need not require a principle transcending the physical world. It differs from both in that its unified view of matter, life and mind avoids the objectionable implications of the two doctrines: an other-worldly realm of mysterious "life forces" animating living organisms on the one hand, and on the other, a cold, meaningless universe in which, through a long series of quite improbable events, molecules randomly combined to somehow generate life and mind. The theory is based on a number of scientific discoveries that, particularly since the 1960s, have been brought together to offer a substantially complete account. Here follows a brief description of its development.

In the early 1900s, certain naturalists (who came to be called organismic biologists) became convinced that a key characteristic of life that neither mechanists nor vitalists adequately addressed was the *relational organization* of living creatures. Individual organisms and communities of organisms display complex configurations of relationships which the biochemist Lawrence Henderson (1878-1942) subsumed under the term "system". From then on, a *system* has come to mean an integrated totality whose essential properties derive not from its constituent parts (as mechanists insist) but from the *relationships* among the parts. These components form multileveled structures of systems within systems, each showing different degrees of complexity and characteristic properties. Organismic biologists were the first to think of life in terms of systems and patterns of relationships.

Interdependence, context, networks, and processes became key notions of the new "systems thinking". That systems are integrated wholes which cannot be understood merely by analyzing their parts is something that was supported by the findings of Quantum Mechanics, the insights of the Gestalt psychologists, and the conclusion of such early ecologists as Arthur G. Tansley, who coined the term "ecosystem" to refer to a community of organisms and their environment.

There was, in other words, a conceptual shift from analysis to synthesis and from parts to whole in studying nature. The properties of wholes exist by virtue of the integrity of an entire system and can only be understood within the context of that system. These properties are destroyed when the whole is dissected or its parts studied in isolation because the essential characteristics of each part arise by virtue of the network of relationships it has with every other part, and the processes they mutually interdepend on.

While systems thinking was slowly making inroads into biology, a new area of science was being established. Drawing from several disciplines and pressed by the urgency of war against the Axis powers, cybernetics (from the Greek word κυβερνήτης, "steersman") was founded in the early 1940s for the purpose of studying problems of communication and control of machines. American Norbert Wiener, one of its principal figures, chose this name for the new discipline because it was recognized that feedback and self-regulation are crucial for the effective function of machines and, it was soon realized, living organisms as well. In other words, communication and control are common features in animals and machines, and the abstract pattern of these shared features is the *feedback loop*.[42]

> A feedback loop is a circular arrangement of causally connected elements, in which an initial cause propagates around the links of the loop, so that each element has an effect on the next, until the last "feeds" back the effect into the first element of the cycle. (Capra: 1996, 56)

One of the most important insights to come out of cybernetics is that feedback loops depict general *patterns*[43] *of organization* inherent in organisms. Both animals and their habitats are links in a dynamic, circular chain of cause and effect in which they are both agents and counteragents. This feedback relationship, according to Wiener, is the essential mechanism of *homeostasis*, the self-regulation and self-organization by means of which organisms maintain themselves in a state of dynamic equilibrium.

By clearly identifying the causal chain that gave rise to the feedback concept, the abstract configuration of any living system's most general behaviour was distinguished from its particular activities and physical structure. An organism, for Norbert Wiener, is essentially a self-regulating entity that is energetically open but organizationally closed. In other words, a living system is an entity that can maintain its pattern of unity even though there is a continuous exchange of matter and energy with its environment.

Thus we see that both organismic biology and cybernetics came to the conclusion that while it is true that all living organisms are composed of atoms and molecules just like all other physical objects, their characteristic difference is certain patterns of organization. The precise nature of these patterns was defined in the early 1970s by Humberto Maturana and Francisco Varela. Before we get to them, some background information.

## *Dissipative Structures and Hypercycles*

The notion that organisms are homestasis-directed self-regulating agents, as we saw, originated in cybernetics[44] during the 1940s. In the 1950s, scientists built binary networks (electronic systems with tiny light bulbs that turn "on" or "off") to model (feedback-sensitive) neural networks. To their astonishment, they discovered that after a short period of random flickering, some ordered pattern (either waves of flickering throughout the network, or repeated cycles) would almost always begin to appear. This spontaneous emergence of order occurred regardless how the initial state of the system was chosen.

It became apparent that in due time, the feedback process of a system imposes a protocol of order that synchronizes random occurrences into certain well-defined patterns. This was a most surprising phenomenon and great efforts were made to find ways to explain them away. But similar observations in many different other disciplines, (including entomology with the study of small social insects such as ants and termites, computerized "cellular automata," mathematical games, in so-called chaotic systems and, as we will see, chemistry) showed that new structures and patterns of behaviours can arise quite naturally out of seemingly random events. This spontaneous, self-induced ordering occurs in dynamic, energetically open systems in which their constituent parts relate in a way that cannot be described in terms of standard linear equations.[45]

> [W]e can say that self-organization is the spontaneous emergence of new structures and new forms of behaviour in open systems far from equilibrium,[vi] characterized by internal feedback loops and described mathematically by nonlinear equations. (Capra: 1996, 85)

One of the most influential and detailed descriptions of self-organizing systems is the theory of *dissipative structures* by the Belgian Ilya Prigogine, for which he won the Nobel prize in chemistry in 1977.

> I wanted to bring together two concepts: the idea of structure, which generally is static;

---

[vi] Equilibrium is a condition in which all influences cancel one another, resulting in a stable, balanced, or static state.

> and dissipation, for which you need energy
> continually brought in and going out.
> (Prigogine: 1983, 17)

Prigogine studied a phenomenon discovered by the Frenchman Henri Benard in the early 1900s, which occurs when a thin layer of liquid is heated from below. As heating increases, the liquid suddenly begins to self-organize, taking on a regular pattern of convection cells. The rotation of the cells alternates horizontally from clockwise to counter-clockwise and, depending on the fluid, square or hexagonal patterns in kaleidoscopic arrangements that resemble miniature stained-glass cathedral windows appear.

> The amazing thing is that each molecule
> knows in some way what the other
> molecules will do at the same time, over
> relatively macroscopic distances. (Prigogine:
> 1983, 13)

This striking example of a self-organization of matter represented, for Prigogine, a significant link between the animate and inanimate worlds that could offer a clue as to how life emerged from non-living matter, for, essentially, life involves chemical reactions that create stability out of instability.

In his careful analysis of the Benard Instability and the Zhabotinsky Reactions,[46] Prigogine showed that as a system moves further and further away from equilibrium, such as, for example, from a state of uniform temperature throughout, it reaches a critical point of instability at which time a spontaneous, self-organized structure suddenly prevails. Non-equilibrium, long considered an unremarkable or even negative state now came to be understood as a source of organization, order and surprising novelty.

> Our [that is, Prigogine and his associates] contribution has been to argue against the idea that equilibrium states are the most important or interesting. On the contrary, it is non-equilibrium that is essential to the understanding of our world and universe. (Prigogine: 1983, 10)

While dissipative structures receive their energy from outside, the instabilities and jumps to new structures of organization are due to fluctuations amplified by internal feedback loops. Each reorganization produces greater complexity that increases the likelihood of random fluctuations and new instability, which impels another transformation to a still more complex organization, and so on. In other words, evolution.[47]

Another Nobel laureate, Manfred Eigen, proposed that the origin of life on Earth[48] may have been the result of just such progressively more complex organizations of chemical systems far from equilibrium and subject to multiple feedback loops. Named "hypercycles," these feedback loops consist of a network of catalysts (molecules that increase the rates of chemical reactions without themselves being altered in the process) in the form of a closed ring. What is especially interesting about these hypercycles is that, not only do they turn out to be stable, but that they are capable of replicating themselves, and of *correcting any errors* while doing so. This intimates the emergence of a mechanism for heredity.

Eigen argues that such self-regulating, self-replicating systems may be a pre-biological phase of evolution — in short, precursors of living systems. For Eigen and others, this is strong evidence that life has its roots in non-living matter.

> The gap between life and nonlife is smaller than we used to believe. Before, we thought that life was the great exception, the contradiction of the laws of physics. Now we see that complexity can spontaneously

arise far from equilibrium. (Prigogine: 1983, 11)

Support for this was the 1952 Miller-Urey experiment. Stanley L. Miller, a student of Nobel Prize-winning chemist Harold C. Urey, performed an experiment that was supposed to model the conditions thought to exist in the earth's early atmosphere. He enclosed methane, ammonia and hydrogen gases inside a loop of glass tubes and flasks. One flask was half-full with water and another flask contained a pair of electrodes. The water was heated to produce water vapour, which mixed with the gases and circulated through the loop to the flask with the electrodes. Sparks were induced to simulate lightning.

At the end of a week's continuous operation, Urey and Miller found substantial amounts of glycine and alanine, two of the thirty-odd amino acids that are the chief components of proteins.[vii] Miller's experiment was the first to show that the building blocks of life could spontaneously form out of non-organic materials.

---

[vii] In 2007, scientists examining the sealed vials from Miller's original experiments found that there were, in fact, well over 20 amino acids necessary for life.

NASA astrobiologist Mike Russell thinks that such processes could have generated life inside iron sulfide deposits in and around ocean floor volcanic vents. The interior of such deposits are permeated with pores measuring hundredths of a millimeter across. These pits would have concentrated hydrothermal fluids, allowing the antecedents of RNA and proteins to form in a sheltered environment, and providing three-dimensional molds for the first cell walls.

So, the ingredients for life appear to be at hand, and the rudiments of structure are provided by hypercycles and iron sulfide bubbles. But there is still something lacking that precludes these from being considered living systems.

Before the 1970s, this missing quality was stubbornly difficult to identify. Biologists would cite characteristics that living organisms share (such as motility, the ability to reproduce, convert one form of energy into another, undergo ontogenic change, etc.) but avoided giving any general definition of life. It was up to Maturana and Varela to provide the essential criterion for clearly identifying what makes a system a *living* system. They named this condition *autopoiesis,* which means "self-making".

## The Systems Theory of Life

The Chilean biologists Humberto Maturana and Francisco Varela began by differentiating "organization" and "structure", two concepts that had often been confused. They defined the *organization* of a living system as the network of relations among its components that determines the system's essential characteristics and demarcates it as belonging to a particular class such as, for example, an amoeba, a coral colony, a kidney. The description of this organization is an abstract map of relationships that has no interest in what the components of the relational network may be.

The *structure* of a living system, by contrast, refers to the actual physical components that make up the system. What is important to keep in mind is that a system's organization is independent of the functional properties of its components. Thus, a given organization can arise in manifold ways by means of completely different kinds of structural components. For example, there are as many as forty types of independently evolved and considerably different configurations of parts that gave rise to a single organizational property: sight.[49]

*Autopoiesis* is a pattern of organization common to all living systems in which the function of each component, however varied it may be, is to participate in the production and/or transformation of other components in order to preserve that system's structural integrity. Thus, the system is both produced by its components and at the same time is involved in the production or preservation of those very components in response to changes in its interior and exterior environments. In short, a viable organism is an integrated network that is continually *making itself*: "In a living system the product of its operation is its own organization."[50]

To define a living organism, then, requires a description of both the system's structure, in terms of physics and chemistry, and a description of its organization, its abstract pattern. Maturana and Varela's characterization of the pattern of life in terms of autopoiesis made it possible for the first time to synthesize the structure-oriented models (such as those of the mechanists, holistic biologists, and of Prigogine) with the organization-oriented models (such as those of the vitalists, cyberneticists and of Eigen) into a coherent Systems Theory of Life. The conceptual glue that binds these two models together is the feedback loop.

Again, a living system, from the simplest cell to entire ecosystems, has a *structure* consisting of physical components, be they organelles, cells, organs, organisms. As physical objects, they are ultimately reducible to molecules, atoms, quarks. In this context, living systems are like non-living things in that they are all equally made of matter. But organisms also consist of certain configurations of relationships among their physical components that mark them out as integrated *animated* units. These patterns of organization are essentially holarchic and, depending on the particular context, they will exhibit different properties that do not exist in other levels, but which are recognizably of living entities.

The patterns of organization are not identical to the physical structures that give rise to them. At the same time they cannot exist unless embodied in those physical structures. In this sense, living systems are no different from machines or artifacts of one type or another. A guitar, for example, is made of different physical components: a sound box, fret board, strings, etc. But it is the relationship among these various components that makes the complex a guitar. It doesn't matter whether the sound box is of wood or metal, or if the strings are of nylon or steel. Provided that all the parts of the structure are there and they are organized in an appropriate pattern, it functions and serves as a guitar.

The difference between a guitar (or any other nonliving system) and a living system is that, whereas in the former the components making up that complex exist *for* each other (i.e. support each other within a functional whole), in the latter the components also exist *by means of* each other (i.e. they *produce* and actively support one another).[51] A living system must continuously monitor and readjust itself structurally to the internal and external environments in order to maintain its organizational integrity. This is what Maturana and Varela meant by "autopoiesis," and it is this disposition which distinguishes living systems from, say, Eigen's hypercycles or Prigogine's dissipative structures. These, too, are organizationally closed and structurally open, but their order and behaviour are imposed by the environment, not autonomously directed.[52]

Living systems can operate far from equilibrium and yet remain integral wholes by responding appropriately and, if necessary, *creatively* to changing circumstances. This remarkable ability suggests intelligence. It is for this reason that Maturana and Varela went on to identify the process of life with cognition, that is, the process of thought.

# PART III

## A NEW VIEW OF MIND

### *The Santiago Theory of Cognition*

It was in the 1970s that Gregory Bateson revived Norbert Wiener's view that organisms can be thought of as self-regulating wholes. Wiener's position had been largely ignored in the scientific community because, with the rise of computer science, von Neumann's model of organisms as information-processing machines appeared to account for animal behaviour reasonably well. But Bateson's study of living systems and the cybernetics debates he was involved in convinced him that the self-regulation which any living organism must necessarily undergo in order to maintain its unity and autonomous entity is fundamentally different than machine processes.

At every level of complexity, the act of living is essentially a *cognitive activity*, a mental process, because every organism, by the very fact that it is a living system, must have both an awareness of its environment and the capacity to respond appropriately to it. It so happens that at the human level, because of our sophisticated brain, the mental process has the uncommon properties of conceptual thinking, language, and self-awareness, but these are not necessary conditions for mind to exist. Mental operations are inevitable attributes of a certain complexity of organization that existed long before organisms developed anything that resembled a central nervous system.

Contrary to the Cartesian-Newtonian model, living systems are not merely automata at the mercy of physical laws, but discerning, purposeful, resourceful, and eminently creative at resolving difficulties they encounter. And we see this at all levels of life, including its outer fringes, extremophiles and viruses. Every living organism, from the simplest to the most complex, must be deemed an intelligent agent in its ability to perceive its environment and take suitable action to maximize its chance of success at staying alive and flourishing. In Bateson's words, "Mind is the essence of being alive." (Capra: 1996, 174)

This radically new view of mind and life was adopted and developed by Maturana and Varela into the Santiago Theory of Cognition. For them also, life and mind were two aspects of a single phenomenon. "Living systems are cognitive systems, and living as a process is a process of cognition."[53]

The environment does not specify or direct an organism's structural changes; it merely provides the conditions for such response. It is the living system, acting with the autonomy of a holon, that not only specifies what structural changes are needed, but also determines which perturbations from the environment will trigger those changes in order to maintain its existential integrity. By specifying which particular disturbances from the environment will trigger particular responses, the system, in Maturana and Valera's words, "brings forth a world."

> Cognition ... is not a representation of an independently existing world, but rather a continual *bringing forth of a world* through the process of living. (Capra: 1996, 267)

Thus, every organism creates a world for itself according to the manner in which it receives and responds to its immediate environment. In other words, every living system constructs its own distinctive ontology according to the particular cognitive structures with which it experiences the world. The experiences themselves will have different qualities of moisture, acidity, salinity, light, temperature, etc. depending on the perceptual capacities of the organism.

And since individual organisms within a species have essentially the same cognitive structure, they bring forth similar worlds.[54] So, while there exists a material world, it does not have any *a priori* features which can be known. It is a known world only potentially, and the depth and richness with which it is understood depends on the receptive and cognitive capacities of the entities that interact with it. Every organism fashions a distinctive cognitive pattern, thereby becoming a psychological centre of a life that is uniquely its own. For example, the world we experience, based largely on our perception of a very narrow band of electromagnetic radiation will be quite different from that of a bee's, whose capacity for sight tends towards the ultraviolet end of the spectrum, or that of a dog's which is largely based on smell.

> The division of the perceived universe into parts and wholes is convenient and may be necessary,[55] but no necessity determines how it should be done. (Bateson: 38)

A single material world for all organisms, then, but necessarily different cognitive models of it based on the different interactions that are had with it. In the process of living and in order to stay alive, all organisms construct unique cognitive topologies of reality, that is, maps.

But, as philosopher Alfred Korzybski pointed out, the map is not the territory: there is more to reality out there than any interpretation can approximate, even our own. Regardless what instruments we use or how we interpret what we experience, what we know about the world is ultimately limited by the make-up of our nervous system that acquaints us with the world, and the conceptual-language structure by which we make sense of it. In short, as circumscribed and incomplete a cognitive model must necessarily be, nonetheless, each and every organism must have one in order to function as a viable organism.

Life is manifestly intelligent, as is evident from the evasion strategies of single-celled organisms, the tactics flowers use to attract pollinators, mimicry in insects, parasitism, symbiosis,[56] and so on. The Santiago Theory of Cognition binds life and mind together in a compelling way.

### *The Gaia Hypothesis*

The identification of life at its most microscopic level with cognition leads to the question as to whether life at the macroscopic level may also be deemed intelligent. A theory that originated in the 1960s and which has received considerable attention in the last few decades proposes this very thing.

While working at NASA in the early days of the space program, biochemist James Lovelock realized that the Earth as a whole displays the characteristics of a living, self-regulating, self-organizing system. He was puzzled by the fact that although the sun's temperature has increased by 25% since life began on Earth, the planet's surface has remained a near constant temperature throughout the nearly four billion years of biotic history. He suspected that life had something to do with this heretofore unaccountable phenomenon but had no idea as to by what mechanism.

In the early 1970s the microbiologist Lynn Margulis provided the answer in the form of various types of *prokaryotes,* i.e. blue-green algae (cyanobacteria) and soil bacteria, that work in the production and removal of gases from the atmosphere. These remarkable creatures are responsible for "fixing" a life-friendly canopy of gases around the planet that, among other things, allows through only certain wavelengths of light which other organisms find useful, while reflecting back into space harmful radiation, including ultraviolet and infrared, that would have made this planet a torrid barren desert. It was the prokaryotes that, according to Margulis, prepared the environment for subsequent and more complex living systems.

Together, Lovelock and Margulis were able to identify a complex web of feedback loops that involved the oceans, rocks, microorganisms, plants and animals in the self-regulation of the planet for the benefit of the biosphere.[57] For them, this phenomenon suggested purposefulness on a global scale. According to what they named the Gaia Hypothesis (after the Greek goddess personifying the Earth), the living and the nonliving constituents of this planet are embraced in a dance of creation which both produces and promotes the conditions for further life. As Lynn Margulis put it,

> Simply stated, the [Gaia] hypothesis says that the surface of the Earth which we've always considered to be the *environment* of life, is really *part* of life. The blanket of air — the troposphere — should be considered a circulatory system, produced and sustained by life … When scientists tell us that life adapts to an essentially passive environment of chemistry, physics, and rocks, they perpetuate a severely distorted view. Life

actually makes and forms and changes the environment to which it adapts. Then that "environment" feeds back on the life that is changing and acting and growing in it. There are constant cyclical interactions. (Capra: 1996, 106)

Life, then, according to this view, is not something that stands apart from or is wholly subject to the environment. Rather, it is its inextricable and essential partner in the generation and proliferation of the myriad living forms on the planet.

If Gaia is the living Earth, then it would be as meaningless to say that life creates its own environments or conditions on Earth as it would be to say that life creates its own environments or conditions in our bodies. If we see Earth as alive, we can still say that its organisms create their environments and are created by them, in exactly the same sense as we say cells create their own environments and are created by them in our bodies. In other words, there is continually and mutually creative interaction between holons and their surrounding holarchies. But we do not divide living bodies or holarchies into "life" and "nonlife." (Harman and Sahtouris: 117)

The Gaia Hypothesis has earned a good deal of interest, especially in light of "deep ecology," a philosophical school founded by the Norwegian Arne Naess in the early 1970s. Whereas one school of ecology (now termed "shallow ecology") is human-centered and considers nature only in terms of its utility to humans, deep ecology recognizes the intrinsic worth of all living things as part of a single biotic community. Human beings are but one node in the web of life that connects every living thing to every other. And, of course, since living things ultimately rely for their livelihood on the non-living world, that is soil, water, atmospheric gases, light, etc., it makes sense to think of organisms and the environment with which they interact as intimately related. For Lovelock, the animate and the inanimate realms form a single continuum: Gaia.

Not only does the biosphere[viii] as a whole evince properties of a living system (defined in terms of

---

[viii] The *biosphere* is defined as a layer of organisms covering, apparently, every square centimetre of the planet's surface. It reaches down as low as 8 km to the bottom of the oceans and extends about the same distance up into the atmosphere. Organisms have been found living in environments of extreme ranges, thriving in the ice of Antarctica, inside rocks, in acid sinks and alkaline lakes, high-pressure vents of superboiling sulphuric springs (as hot as $350^{\circ}C$), lakebeds ten times saltier than the oceans, and in ground water 3.5 kms below the surface of the earth. The *Deinococcus radiodurans* bacterium can survive a

autopoiesis, structure and cognition), but it also satisfies what the biologist Gail Fleischaker has identified as the three criteria of all autopoietic networks: self-boundedness, self-generation and self-perpetuation. For example, the Earth's atmosphere was created and is now being maintained by the biosphere's metabolic processes. It is semi-permeable like a cell membrane, only allowing in certain radiation, keeping gases from escaping, etc. and, this being so, it can be said that Gaia is *self-bounded*. It is also *self-generating* in that all the components of the Gaia network process organic and inorganic substances into means that promote its continuation. And it is also *self-perpetuating* because the favourable but dynamic conditions it creates lead to the evolution of novel responses and new organisms.[58] And, of course, if humans (or some intelligent, silicon-based creatures we will develop into or merge with) ever get to colonize other planets, then Gaia's self-perpetuation will have achieved a new level of fruition.

---

15,000-gray dose of radiation, whereas 10 grays would be fatal to a human. (It takes 1,000 grays to kill a cockroach).

Life has a tenacious hold on this planet. An ecological or nuclear disaster may prove to be catastrophic for humanity and assorted other species, but life in some capacity will carry on, at least as long as the sun continues to burn, if not longer (see endnote 121).

In this light, the Earth resembles, as the science writer Lewis Thomas has pointed out, nothing so much as a self-contained cell. The various ecosystems constitute an autopoietic network (analogous to organelles) that monitors its internal and external environments and maintains itself as an integrated holon. The biosphere and the environment make up a single living entity that has not only survived four billion years of catastrophe and mass extinctions, but has evolved into an immensely rich variety of which we know only a tiny fraction. The evolution of living organisms is so closely tied to that of the environment that a single evolutionary process operates for both. This is also true within particular ecosystems.

> Rabbits evolve together with their "rhabitats," so to speak — all creatures evolving in connection with all else evolving around them. It took a century and more after Darwin's theory was published for us to understand that environments are not ready-made places that force their inhabitants to adapt to them, but ecosystems created of, by, and for living beings ... as they transform and recycle the materials of the Earth's crust. (Harman and Sahbouris: 63)

## *An Intelligent Universe?*

Intelligent interaction between organisms and the environment occurs at every level of the holarchy. Returning once again to the simplest organisms on earth, the non-nucleated prokaryotes, we have discovered in the last three decades that they are astonishingly resourceful. They are, by far, the most successful organisms on the planet, occupying every known habitat, from the moderate to the most extreme. They, for example, cover every surface of our body, external and internal, from the skin to the inner linings of our sinuses, esophagus, stomach, intestines, outnumbering our own 50+ trillion cells at least ten to one. But what is perhaps most remarkable is that they form, in their totality, a single planet-wide living system spanning the entire biosphere and in which, it can be said, all other organisms live and move and have their being.

> There they are, moving about in my cytoplasm ... They are much less closely related to me than to each other and to the free-living bacteria out under the hill. They feel like strangers, but the thought comes that the same creatures, precisely the same, are out there in the cells of seagulls, and

whales, and dune grass, and seaweed, and hermit crabs, and farther inland in the leaves of the beech in my backyard, and in the family of skunks beneath the back fence, and even in that fly on the window. Through them, I am connected: I have close relatives, once removed, all over the place. (Thomas: 86)

Beneath our superficial differences we are all of us walking communities of bacteria. The world shimmers, a pointillist landscape made of tiny living beings. (Lynn Margulis and Dorion Sagan, as quoted in Capra: 1996, 239)

The prokaryotes have access to a single gene pool and can restructure their genomes in response to stress by transferring genetic material among each other "in a global exchange network of incredible power and efficiency."[59] Resistance to a particular antibiotic can be passed from bacterium to bacterium and incorporated into the genome of the entire global bacteria population within a matter of days. We are presently using fifth generation antibiotics because the targeted bacteria have evolved new strains in response to the threat. Viruses,[60] as well, have responded the same way to antiviral drugs.

And, as the experiments of molecular biologists John Cairns and Barry Hall in the 1980's showed, they can apparently direct their own evolution by generating useful mutations many millions of times more frequently than normal (i.e. randomly) in response to a hostile environment.

Certainly for many, the way in which bacteria interact with their surroundings, with each other, and with higher-level holons suggests purpose and, therefore, intelligence. But if intelligence is found at so simple a level in the holarchy, the question arises as to whether it is an emergent property of life or whether it is logically or temporally prior to life. A controversial view that has sparked heated debate in the last decades argues that intelligence is an inherent aspect of *matter* itself. In other words, consciousness pervades the entire universe and thus, the universe must itself be considered intelligent/alive. This idea, Panpsychism, has a long history stemming back to animist religions, mystical traditions, and venerable idealist philosophies. It has been given new life in recent decades, however, not only in light of the findings of Quantum Mechanics, but because of the abject failure of Materialism to resolve the so-called "hard problem" of how to account for the indisputable fact of consciousness.

The bizarre nature of the subatomic realm, and the impossibility of separating mind from the physical world in trying to explain it, as previously mentioned, has convinced many that such a quasi-idealist, or outright idealist metaphysics (from Advaita to Bernardo Kastrup's), offers the best account of Reality. The universe resembles a great thought far more than it does a giant machine, said the British astronomer James Jean.[61] The evolutionary biologist Elisabet Sahtouris concurs with this view.

> I personally believe that consciousness does indeed permeate the universe, that the universe proceeds intelligently in its evolution and must therefore be conscious.[62]
> (Harman and Sahtouris: 123)

So did the great physicist Max Planck.

> I regard consciousness as fundamental. I regard matter as derivative from consciousness. We cannot get behind consciousness. Everything that we talk about, everything that we regard as existing, postulates consciousness.[63]

Regardless which level of the holarchy we choose to examine, we find holons involved in perpetual dynamic interaction with other holons. Energy and matter are drawn in, used, transformed, exchanged, and not only in living organisms. The entire universe, from the subatomic realm to planetary systems to galaxies to galactic supergroups, displays a supremely complex net of interconnections that, for many, strongly suggests purpose.[64]

> We can find intentional, goal-oriented, or teleological behaviour wherever we look in nature. (Harman and Sahtouris: 35)

> The interlocking web of evolving nature is not blind or chance-like but becomes, as it progresses, rich in irreversible, directional, ever more complex constraints that tend to keep things moving in a trend toward higher and more competent forms. (Sperry: 1985, 118)

If intelligence, purpose, or consciousness is manifest in any level of the holarchies of nature, then it must necessarily be a characteristic of it in its entirety. Again, Sahtouris:

> If we adopt the model of the universe as a living evolutionary process generating such holarchies, this conception of self-transcendence as conscious intelligence

> begins to give us a very exciting "radical unity." What is so important is that this model gives us a way of talking about consciousness and intelligence not as an emergent property of evolution ... [but] as a fundamental feature of every holon of a living universe from its inception. (Harman and Sahtouris: 126)

The view that the universe is "alive" or "conscious" is presently accepted by a small but growing minority among the scientific community, having won the support of some highly-respected thinkers, including Carl Sagan, John Wheeler, David Chalmers, Thomas Nagel, Galen Strawson, et al. The reason is that, along with the evidence of what seems to be purpose throughout the physical world, this hypothesis would resolve a number of perennial conundrums that not only includes the origins of intelligence, life, and consciousness, but also the universe's evolution, its fitness for life, as well as the mind/body problem, free will vs. determinism, and experiences reported during altered states of consciousness (e.g. mystical union).

To argue that life, intelligence and self-aware consciousness are emergent properties of sufficiently complex holons does not answer the question as to *how* such could come about. If, on the other hand, intelligence is a primary, essential aspect of the cosmos from the very beginning ("a fundamental feature of every holon of a living universe from its inception" — Sahtouris) then physical laws, evolution, life, self-awareness, culture, and so on could be explained in terms of facets or manifestations of this intelligence. In the words of systems theorist Margaret Wheatley, "We live in a universe that is alive, creative, and experimenting all the time to discover what's possible."[65]

The properties that emerge in the unfoldment of the universe over time emerge by virtue of an inherently creative and purposive Spirit suffusing the universe.

> Spirit is Nature and Nature is Spirit, if you like those terms, though the concepts, not the labels, are important here. (Harman and Sahtouris: 174)

This position resonates with a recent theory known as Biocentrism. According to its chief proponent, biologist Robert Lanza, life is the source of being in that it created the universe. Space, time, and all objects and processes within them are mental constructs and not entities with any "objective" reality. Life alone is axiomatically real. In the words of the Indo-American endocrinologist and writer Deepak Chopra, who supports Lanza's view, "Consciousness preceded the brain: it created life and went on to create the brain itself."[66]

Lanza's view is that life, or more precisely consciousness, is necessary for the universe to be. In other words, the universe could not exist unless witnessed by some conscious observer.[67] Space, time, matter, energy, forces, … *everything* that is discerned as "being" is created by the act of perception.

Like other Idealist-tending "theories", Biocentrism is interesting and seemingly on the right track, but it is not sufficiently robust to be satisfying. It does not give an adequate explanation of *how* the act of perceiving can create the *very conditions* of what is experienced. Nevertheless, its basic tenet, that life/intelligence is central to a comprehensive understanding of the universe is a point that many scientists, including Sahtouris, sympathize with.

It would explain, for example, how the phenomenon of life, which tends towards ever greater order, complexity and diversity can so blatantly violate the Second Law of Thermodynamics, which dictates that the universe tend inexorably toward decay, entropy, absolute *dis*order:

intelligence is behind both the phenomenon of life and the Second Law of Thermodynamics. It has transformed itself into energy and then into matter, and now uses that matter to express itself as living organisms.[68]

For some, such idealist accounts smack of the occult for, like Vitalism, they require some invisible principle behind physical phenomena. An alternative to both the idealist and materialist explanations of the universe has been proposed by astronomer George Greenstein.

### *The Symbiotic Universe*

According to Greenstein, two facts must be reconciled.

> [F]irst, the fitness of the cosmic environment for life, a fitness that hangs by a thread and that cries out for explanation; and second, Quantum Mechanics, with its extraordinary insight into the role played by observation in the creation of reality. (Greenstein, 240)

In Quantum Mechanics, Schröedinger's wave equation $\psi$ is used to represent how a quantum system evolves through time. (The function plays a role analogous to that of trajectories in classical mechanics). However, although the equation predicts a smooth evolution of the quantum system over time, it can give only a probabilistic account of a particular state at any one moment. As previously mentioned, whether an event occurs or does not occur cannot be determined with any definite certitude, and the only way to have any knowledge at all of its state is through observation. In the words of Nobel physicist John Wheeler, "No phenomenon is a real phenomenon until it is an observed phenomenon."

What we take for the world, then, is something that comes about through the act of perceiving. Whether an electron, say, is a particle or a wave depends on the instrument one chooses to measure it with. If one sets about to measure an electron's frequency, then the electron will show its wave character. If, on the other hand, one wants to measure its mass or spin, then it will reveal its particle side. Before the observation is made, an electron is neither particle nor wave but a "superposition" of the two states, that is, solid matter and immaterial motion simultaneously. What is "out there," then, is undifferentiated ur-stuff that depends for its existence as distinct objects, relationships, processes, etc. on the categorizing ability of mind to assign it distinguishable qualities.[69]

According to Nobel laureate Eugene Wigner and others, consciousness alone is not subject to the probabilistic incertitude inherent in Quantum Mechanics, which, they point out, is not restricted to the subatomic realm: not only is there no line of demarcation between the quantum world and the macro-world,[70] but it has been shown that even macro-world objects such as buildings, planets and galaxies are subject to Schröedinger's equation. If consciousness is not dependent on $\psi$, then the implication is that mind is somehow outside the purview of physical law. But this is highly unlikely, as evidenced by the effects that drugs, hormones, and brain injury (e.g. the strange and sad case of Phineas Gage) have on mental states. Clearly, chemistry, and by extension physics, have some role in mental events.

If neither an idealist nor a materialist metaphysics (whereby life and mind are epiphenomena of physical processes) can adequately account for our experience of reality then, according to Greenstein, one explanation remains: there must exist a *symbiotic* relationship between mind and matter. Mind requires matter in order to become grounded in some physical entity (a disembodied mind is untenable), and matter requires mind to become real (again, without the conscious perception of some mind, an object remains an unqualified entity real only in potential). The mental and the material, therefore, are not two separate realms but intimately interwoven, constantly informing and shaping each other as a single ontological entity. To use a trite but fitting metaphor, mind and matter are two sides of the same coin: without the other, neither could exist.

> The very cosmos itself depends for its being on the uttermost mystery of consciousness. And thus the symbiosis, the union between the physical world and mind, the great metaphysical dance by which each brings into being the other. (Greenstein 238)

Greenstein's view, like Lanza's, interesting as they may be, can't be considered "scientific" theories because they are not falsifiable and because they do not make any testable predictions, at least, as of yet. They can be thought of as interesting and even compelling metaphysical hypotheses, but that is all we can expect of them for the time being. Nevertheless, they should be given due consideration because they are our best attempts at trying to make sense of a universe that is far more mysterious than was ever imagined. In so doing, and as feeble as the results appear to be, we are very likely approaching the limits of what may be comprehensible[71] to the human species.

# PART IV

## A NEW SOURCE OF VALUES

### *E Duobus Unum*

Regardless whether intelligence is an integral aspect of the universe at its most fundamental as Lanza, Sahtouris, and Greenstein separately hold, an emergent property of matter (Bateson), a facet of the *élan vital* (Henri Bergson), or a creative principle (Alfred North Whitehead) permeating the universe yet separate from matter (Arthur Eddington and George Wald), the primacy[72] of mind is another foundational notion of the New Science.

As we saw, Relativity Theory and Quantum Mechanics differ radically from Cartesian-Newtonian physics in that they recognize that mind plays an essential role in any experience and, hence, in any description of reality. Strangely enough, this fact was not acknowledged in psychology until the 1960s. Before that time, Behaviourism, with its emphasis on the observable actions of organisms to the exclusion of introspective data or any reference to mind and consciousness, had near hegemonic sway over the discipline. But studies involving psychedelic drugs, especially LSD-25, led many psychologists to question the behaviourist doctrine and began to scientifically reconsider subjective experience.[73]

At about the same time, Roger Sperry was performing important experiments (for which he eventually won the Nobel Prize) with epileptic patients whose *corpus callosum*, the main nerve connecting the two halves of the brain, had been severed. What he discovered was that when this connection was cut, the patient behaved as though there were two minds inside one brain, with each hemisphere perceiving, learning, remembering, and feeling things that were completely unknown to the other half.

> We found that each disconnected hemisphere was capable of sustaining its own conscious awareness, each largely oblivious of experiences of the other. The separated hemispheres were able to carry on independently at a fairly high level. They could even perform mutually contradictory tasks at the same time, and each was able to exert its own volitional control and select its own differential preferences. (Sperry: 1983, 74)

The right hemisphere, it soon became clear, experiences reality in all its complex wholeness (shape, colour, motion, sound, smell, taste, touch, emotional response, etc.) moment by moment in an ever-present now. The left hemisphere takes each of these momentary slices of experience and strings them together into a chronological, inferential sequence that structures reality and "makes sense" of it. In short, while the left hemisphere is highly verbal, mathematical, and operates in a linear, coldly-analytic manner, the right hemisphere is mute, all-embracing, and operates in an intuitive, holistic, emotionally-qualified way.

In one celebrated experiment, blindfolded patients were asked to sort, by touch, differently shaped beads. One hand would sort out spheres into an upper tray and cylinders into a lower tray, while the other hand would do the opposite. In the process of sorting, each hemisphere would consciously and voluntarily make decisions contrary to those being made by the other hemisphere, without either knowing what the other was doing. The verbal left hemisphere could give a moment by moment report of what the right[74] hand was doing, but had no clue as to what the left hand was doing or whether it was moving at all. But if each hemisphere of the brain has its own conscious-volitional field, why is it that in the normal state we experience a single[75] unified mind? Sperry concluded,

> The normal bilateral consciousness can be viewed as a higher emergent entity that is more than just the sum of its right and left awareness and supersedes this as a directive force in our thoughts and actions. (Sperry: 1983, 74)

In other words, when the two hemispheres are connected and function normally, there is a sense of one consciousness because some prevailing entity arbitrates and integrates the chemical activities within each hemisphere and translates them into one coherent experience. This authoritative principle, the single mind, is a higher-level holon relative to the neuronal network in the left and right hemispheres of the brain that gave rise to it. The emergent properties that go with the single mind not only differ qualitatively from those of its components but are also able to exercise prerogative over those components.

> I have used the example of how a wheel rolling downhill carries its atoms and molecules through a course in time and space and to a fate determined by the overall system properties of the wheel as a whole

and regardless of the inclination of the individual atoms and molecules. The atoms and molecules are caught up and overpowered by the higher properties of the whole. One can compare the rolling wheel to an ongoing brain process or a progressing train of thought in which the overall properties of the brain process, as a coherent organizational entity, determine the timing and spacing of the firing patterns within its neuronal infrastructure. The control works both ways; hence mind-brain "interaction." The subsystem components determine collectively the properties of the whole at each level and these in turn determine the time-space course and other relational properties of the components. (Sperry: 1985, 94)

Thus, for Sperry, human consciousness is neither a pseudo-problem to be resolved through an analytic-linguistic approach, nor an epiphenomenon of biochemical activity along neuronal pathways in the brain. Rather, it is a genuinely real phenomenon, a "causally potent reality" that both directs and is a product of the biochemical processes of "an evolving, self-creating cerebral system" and inseparable from it. In short, mind and brain co-exist as truly autonomous entities, but in a mutually dependent unity.

## *The Transpersonal Mind*

Another current that led to the so-called humanist or mentalist revolution[76] of the 1970s, whereby the study of consciousness, once considered a lowly adjunct to the physical sciences, suddenly became the centre of attention in science, was the work of such clinical psychologists as the Czech Stanislav Grof. One of the first scientists involved in studying the effects of psychotropic (mind-altering) drugs and the non-ordinary states of consciousness they induce, he became convinced that "the traditional understanding of the human personality, limited to postnatal biography and to the Freudian individual unconscious, is painfully narrow and superficial." (Grof: 20)

Although he had begun his professional career as "a convinced materialist and atheist," the remarkable intensity and similarity of experiences in altered states (including his own, as he was obliged to participate personally in the experiments in order to understand the phenomena "from the inside," another feature of the New Science) with those reported in mystical and shamanic traditions led him to conclude that the experiences were not the result of a psychosis brought on by the drug, but a glimpse of a world beyond our everyday reality. He believes that there exists a "transpersonal" realm to which we can have access under certain special circumstances.

While the commonplace notion that human consciousness is contained within our individual skulls may appear true with respect to everyday states of consciousness, it cannot explain what happens when we enter non-ordinary states such as trances, spontaneous psycho-spiritual crises, or those attained through hypnosis, psychedelic sessions, meditation, and experimental psychotherapy.

> When we enter the realm of transpersonal experience we burst through barriers that we take completely for granted in our everyday lives. At this point various historical events, moments that belong to the future, and elements of the world that we would normally consider to be outside the range of our consciousness, appear to be as real and authentic as anything we have ever experienced. We can no longer assume that what we encounter here are products of our imaginations. The world of the transpersonal exists quite independent of us. (Grof: 86-87)

Experiences of the transpersonal realm led to therapeutic results that exceeded anything that had been theretofore tried. Symptoms that had resisted months or even years of psychoanalysis and other treatment often vanished after patients underwent but a few of these kinds of sessions.[77] (Grof: 17)

Grof's studies, like Sperry's, confirmed the primacy of experience, the irreducible nature of consciousness, and the mind's causal efficacy over matter. Grof's interpretation of his findings led him to adopt a metaphysics very much like that of Sahtouris.

> I see consciousness and the human psyche as expressions and reflections of a cosmic intelligence that permeates the entire universe and all of existence. We are not just highly evolved animals with biological computers embedded inside our skulls; we are also fields of consciousness without limits, transcending time, space, matter, and linear causality. (Grof: 17-18)

Like the late Canadian neurosurgeon Wilder Penfield, he believes that consciousness does not have its source in the brain.[78] Rather, it is "mediated" by the brain in a way analogous to how radio signals are picked up by suitably-tuned radios. Thoughts arise from an infinite field of consciousness beyond time and space. They become coherent only according to conformable receivers, but the signals are always there, suffusing space-time.

> [O]ur individual psyches are, in the last analysis, a manifestation of cosmic consciousness and intelligence that flows through all existence. We never completely lose contact with this cosmic consciousness because we are never fully separated from it. (Grof: 202-203)

This view is much like the one held by the world's mystical tradition, what Aldous Huxley referred to as the "perennial philosophy."[79] Central to these ancient doctrines is the belief that material things are somehow embedded in a living universe which, in turn, is a manifestation of consciousness or spirit. Everything consists of a continuum of being, at one end of which is insentient matter and at the other end Spirit or superconsciousness, which also happens to be the all-pervading ground[80] of all that exists. Everything is interconnected and constitutes a Unity, both physically and psychically, but mind/spirit is the fundamental essence: matter is but a category of this substratum, an expression of its creative impulse. (Harman and Sahtouris: 192-193)

Whereas Grof's position tends to an idealist monism, Sperry's is firmly grounded on a materialism that views consciousness and all mental phenomena, including animal awareness and behaviour itself, as arising out of but not reducible to the physiochemical processes of the brain.

> As an emergent functional attribute of brain activity, conscious experience is inextricably linked to, and inseparable from, the functioning brain. It is only in the functional relations within the matrix of brain processing that the subjective qualities appear and have their meaning. The subjective effects are generated by, and exist only by virtue of, brain activity. Even where higher order mental forms are compounded of lower level mental entities, as we assume to be the case, the entire hierarchy is still embodied in, dependent on, and inseparable from the physiological substructure. (Sperry: 1985, 98)

Between these two opinions, sits the long-dormant doctrine of dualism, which made a surprising comeback in the late 1970s when Karl Popper and another renowned authority in science and philosophy, John Eccles, jointly published *The Self and its Brain: An Argument for Interactionism.* Creations of the mind,[81] they argued, such as myths, music, abstractions, witticisms, mathematical formulas, etc. cannot be accounted for by the laws of physics or physiology. At the same time, the brute fact of the physical world and the close correlation between the physiological and psychological states (the presence of certain hormones whenever pleasure or pain are experienced, the effects of narcotics, the efficacy of drugs for dealing with psychological problems, etc.) suggest that mental states are reducible to biochemical processes and, therefore, to physical processes. Consequently, there must exist two separate realms, a material and an ideal. They "interact" but remain essentially separate.

Most scientists and philosophers dismiss dualism because it fails Occam's test for economy in explanation, and because it does not resolve a basic difficulty with dualism: what is the interface between such radically different realms as mind and matter that would allow them to interact as they obviously do? Sperry, himself, has felt the need to refute Popper and Eccles, especially since they cite his work as supporting their views. He is emphatically monistic.

Mentalism is strictly a one-world, this-world answer. I don't see any way for consciousness to emerge or be generated apart from a functioning brain. Everything indicates that the human mind and consciousness are inseparable attributes of an evolving, self-creating cerebral system. (Sperry: 1983, 75)

### *Science and Values*

Perhaps the most important implication of the mentalist revolution, according to Sperry, is that we can now have a much more comprehensive world view, one which acknowledges both our interior and exterior realities.[82] As British philosopher Owen Barfield has said,

[T]here is indeed only one world, though with both an inside and an outside to it, only one world experienced by our senses from without, and by our consciousness from within. (Harman and Sahtouris: 159)

Sperry argues that we can integrate the reductionist, exterior-directed view of scientific materialism with the holistic, interior-directed view of the humanist philosopher. In short, Kant's "starry heavens above" and "moral law within" can be reconciled in terms of a single phenomenon: mind.

In the past, philosophy taught that science and values were separate, that science *de*scribes but cannot *pre*scribe, that science can tell us *how* but not *why*, that it can tell us what *is* but not what *ought* to be, that it can achieve objectives but cannot tell us which goals are worth pursuing, etc. With respect to conflicts in values, we were told to seek answers elsewhere, be it the humanities, ethics, religion. Nor could science help us to cope with the social neuroses of our time, which include a sense of hopelessness and a loss of purpose or higher meaning. In short, science had nothing to say regarding meaning, duty, personal worth, dignity, and the like — values that are central to human life.   With our new understanding of the mind in terms of causal connections, the traditional separation of science and values is no longer valid. Since all we can experience or know is a product of the mind, values are no longer isolated, incontestable principles, but causally connected to all other mental creations and, hence, open to scientific treatment.

> Subjective values, like other mental phenomena … can be treated in principle, along with facts, as causal agents in brain

processing and in the objective world, and thus are a legitimate concern of objective science. (Sperry: 1985, 15)

According to Sperry, consciousness, as is experimentally measured by fMRI, encephalographs, etc. can exercise causal authority over brain activity. Thus, we have a measure of true autonomy over what we think, intent, and do — autonomy, of course, that is circumscribed by our position in the holarchy of existence. We are not absolutely free because we are subject to the constraints of both the components that constitute us, and the outer holons of which we are part. In this sense, we are determined by spatio-temporal circumstances outside our control. But insofar as we can will ourselves to act upon and respond to these constraints according to the plans we conceive or a future we imagine, we are free.

[Actions] are free to an extent. We're no longer subject to, or in the grip of, the laws of physics and chemistry, as inanimate objects are. Nor do we have to obey the laws of physiology, as do our autonomic and reflex responses, our hormones, and our heartbeat. In general we are free of the kind of mechanistic materialist forces with which science used to saddle us. We are lifted above these into a higher realm with a different kind of control — a control unequaled in freedom anywhere else in the known universe. (Sperry: 1983, 75)

The ancient philosophical problem of free will vs determinism is resolved: we are both determined and free. This conclusion, for Sperry, cannot be overestimated, for it can now be said that we are truly responsible for our own actions and behaviour. We are no longer puppets of biochemical processes, genetic programs, instinct, subconscious motives, or our personal histories. While these may have some influence over our behaviour, our conscious thoughts can supersede them: we can always *choose* and will ourselves to do otherwise, and choice is ultimately self-dependent and personal.

Any given brain will respond differently to the same input and will tend to process the same information into quite diverse behavioral channels depending on its particular system of value priorities. In short, what an individual or a society values, determines very largely what it does. (Sperry: 1985, 109)

Our actions are guided by our values, and because we are, to a large extent autonomous, we can create our values, not from a sense of angst, as Existentialism warrants, but with forethought and deliberation of ends we desire and choose. This insight is particularly important today because our value priorities are now effectively the most powerful causal agents on the planet (barring the occasional slap into temporary humility delivered via tsunamis, wildfires, hurricanes, pandemics, etc.)

> The outstanding feature of our times is the occurrence of [a] radical shift in biospheric controls away from the vast interwoven matrix of pluralistic, time-tested checks and balances of nature, to the much more arbitrary, monistic, and relatively untested mental capacities and impulses of the human brain. (Sperry: 1985, 8)

The forces of nature that have regulated events in our biosphere for nearly four billion years have now quite abruptly been superseded by the actions of our species. Consequently, in Sperry's words, "there is no more critical task" at hand than establishing a new, more enlightened set of value guidelines that will guide us in dealing properly with our world so as to protect and preserve the biotic community instead of continuing to harm it and, perhaps, sowing the seeds of our own destruction in the process. There is need for a global ethic that is accepted and respected by all nations as the principal standard for judging ethical priorities. This ultimate criterion will be based not on outdated, partisan, or otherworldly frames of reference according to which humanity has traditionally tried to live by, but on reason and the rigorous standards of science.

> In other words, founded in a reference framework based on empirical evidence and the scientific method as the best avenue to truth available to the human brain and also on the world view that derives from science, i.e., the world model and view of reality supported by the total collective knowledge of all the sciences along with the insights and perspectives this knowledge brings. (Sperry: 1985, 46)

Sperry makes clear, however, that the science he is referring to is the New Science, which "stands for something that is no longer in conflict with ethical, religious or other humanist sensitivities." (Sperry: 1985, 47)

This new source of values, he argues, can provide the foundations for a new ethic, something that neither the

Cartesian-Newtonian view nor traditional religion can any longer offer credible grounds for. The result, he admits, will not alter most quotidian conduct or morality, but it would lead to significant changes with respect to how we deal with the natural world. There would be greater appreciation for the biodiversity we see all around us, and greater concern for its welfare. It would lead to an ideology

> … that would make it sacrilegious to deplete natural resources, to pollute the environment, to overpopulate, to erase or degrade other species, or to otherwise destroy, demean, or defile the evolving quality of the biosphere. (Sperry: 1985, 115)

Sperry's point is that, now that we are the major causal agents in nature, it is up to us to be proactive in protecting and sustaining it. This requires a radical reconsideration of our values and our understanding of the world, founded on "New Science" thinking.

# PART V

# A NEW PARADIGM

## *Today's Crisis: Pathosis of a Misperception*

Every age has had its defining problems, be they warfare, famine, disease, social unrest, or decadence. Today's ecological and social problems are particularly daunting for a number of reasons. First, they are numerous and interrelated; they cannot be understood let alone be solved in isolation from one another. Second, they are global in scale, and affect not only human beings but the entire biotic community for all future generations. Third, our political, cultural, and business leaders lack either the power or the will to implement the necessary large-scale changes.[83] ("The American way of life is not negotiable," said President George Bush at the 1992 Earth summit in Rio.)[84] Fourth, to solve these problems will be costly.[85]

Citizens of the developed nations are not only unwilling to give up their lifestyles, but are openly encouraged, through advertising,[86] to evermore, often outrageous consumption (witness the popularity of fuel-guzzling SUV's in the decade following the Rio summit).

The rest of the world's people (the poorest 20% of the world's population earn less than 2% of the world's income)[87] meanwhile, have more pressing priorities, such as feeding their children today than worrying about how destructive their agricultural or industrial methods may be in the long term. When the richest nations of the world refuse to curtail their consumption, which is depleting the planet's resources many times faster than all the developing nations combined, who are they to demand that the poorer nations forego their own opportunity for material comfort? And so, the problems continue to worsen, becoming evermore intractable.

What is not commonly understood in our current state of affairs is that all the problems mentioned are aspects of a single crisis, a crisis that Capra, Sperry, and the cultural historian Thomas Berry, among others, consider a crisis of understanding, of trying to live without a sense of "cosmic significance."[88] As the great sociologist Max Weber put it, the modern world is "disenchanted". It has been sanitized of any spiritual, symbolic or expressive dimension that provides a cosmic order in which existence finds its meaning and purpose. We plod along our directionless, destructive way because we lack an appropriate conceptual framework by which we can get our bearings.

> It's all a question of story. We are in trouble just now because we do not have a good story. We are in between stories. The old story, the account of how the world came to be and how we fit into it is no longer effective. Yet we have not learned the new story. Our traditional story of the universe sustained us for a long period of time. It shaped our emotional attitudes, provided us with life purposes, and energized action. It consecrated suffering, and integrated knowledge. We awoke in the morning and knew where we were. We could answer the questions of our children. We could identify crime, punish transgressors. Everything was taken care of because the story was there. It did not necessarily make people good, nor did it take away the pains and stupidities of life, or make for unfailing warmth in human association. It did provide a context in which life could function in a meaningful manner. (Berry: 123)

Our current condition is due to the fact that most of us, and especially our decision makers and influencers, still subscribe to an outdated worldview based on a narrative that is no longer either valid or reliable for providing direction as to how to deal with our overpopulated, economically unbalanced, resource-insatiable world. There are solutions to the critical problems before us, and the world community can mobilize to solve them, if it so chooses. But before we can will ourselves to do so, there must be a fundamental shift in our perceptions, our way of thinking, our values.

### *The Ecosophical Paradigm*

The new perspective this conceptual and ethical shift requires is a corollary of the global mind change that Michael Thoms heralded in the first pages of this book. A prominent expositor of the emerging mindset is Fritjof Capra who, in books, articles and lectures, has done much to elucidate and promote it. He calls this present time of transition between the old and the new world views the "Turning Point".

> The new vision of reality we have been talking about is based on awareness of the essential interrelatedness and interdependence of all phenomena — physical, biological, psychological, social, and cultural. It transcends current disciplinary and conceptual boundaries and

will be pursued within new institutions. At present there is no well-established framework, either conceptual or institutional, that would accommodate the formulation of the new paradigm, but the outlines of such a framework are already being shaped by many individuals, communities, and networks that are developing new ways of thinking and organizing themselves according to new principles. (Capra: 1982, 265)

It is the synthesis of the New Science and environmentalism that constitutes the new world view, which I have named the Ecosophical Paradigm.[89]

The term "Ecosophical" is derived from the Greek words οίκος, meaning "household" or "home" and σοψία, meaning "wisdom". Although not a neologism,[90] it is not a familiar term and I am adopting it for our purpose here because I think it best labels the emerging *Weltanschauung*: In the past half-century, we have become not only more knowledgeable about our home, the Earth, and the creatures with whom we share it, but wiser too (or so one hopes). As well, we are apprehending more and more our planet's own inherent wisdom. The new world view acknowledges both these kinds of wisdom and, for this reason, I think

"Ecosophical" is an apt term. Now a word about "Paradigm".

When Thomas Kuhn referred to "paradigm shift" in his *The Structure of Scientific Revolutions* (1962), little did he anticipate the large influence the term would turn out to have. In fact, along with "relativity" and "quantum", "paradigm" has become one of the most widely appropriated scientific terms of this century, used in so many different senses than Kuhn originally intended[91] that he himself later abandoned it.

According to Kuhn, scientific progress is not incremental but proceeds in stages via a *paradigm shift*, in which a complementary bundle of accepted theories that reflect and uphold certain established viewpoints is replaced by another. This is because in the course of its development and refinement, a scientific theory will encounter anomalies. When too many exceptions to the theory obtain, however, competing explanations gain stature. Since the reigning paradigm is often part of a larger world-view, is supported by tradition, and careers depend on it, it resists the force of an alternative position. A period of instability ensues in which the old and new paradigms compete for acceptance by the scientific community. Eventually, and usually after the death of eminent supporters of the old view,[92] the new view becomes the dominant paradigm until it too is usurped by a newer one. The study of science, then, is not a neat logical accretion of discrete theories and discoveries, but a never-ending battle royal between competing explanations of ostensibly the same observations, in order to shake out a consensus, albeit a temporary one, among the scientific elite.

For Kuhn, almost any experiment generating significant data could be deemed a "new paradigm". Hence, in the history of science, Kuhn saw not just a handful of paradigm shifts, as is often believed, but hundreds.

Notwithstanding Kuhn's original use of the word, in common parlance "paradigm shift" has come to represent a decisive displacement of some overarching vision of reality by another comprehensive set of concepts, values, etc. It is in this sense that I use the term because the emerging Ecosophical Paradigm shift is considered by many thinkers (including Capra, Harman, and Berry) to be indeed one of the half-dozen or so major cultural transformation that have occurred in the history of humanity. Certainly, insofar as its champions claim, it is at least as momentous as Copernicus' heliocentric cosmology or Darwin's evolutionary perspective.

## *The Characteristics of the New Paradigm*

There are different ways[93] to characterize the change in perspective the Ecosophical Paradigm proffers. One is derived from the work of Capra and Steindl-Rast, who have identified five features of what they call New Paradigm thinking. Here is a summary.

1) *A Shift from Part to Whole*. Whereas in the Cartesian-Newtonian view, the dynamics of the whole were to be understood in terms of the properties of its constituent parts, in the new paradigm, the properties of the parts can be understood only from the dynamics of the whole. As mentioned, the ultimate development of this view is "bootstrap theory," whereby there are no fundamentals or even parts. What we would call, say, a bird, or a tree, is merely an abstraction from an essentially inseparable web of relationships.

2) *A Shift from Structure to Process*. Whereas in the old paradigm, it was thought that there were elementary structures, and then forces and mechanisms by means of which these structures

interacted, in the new paradigm, every structure is considered a manifestation, not an agent, of underlying processes.

3) *A Shift from Objective Science to Epistemic Science.* In the old paradigm, scientific descriptions were deemed to be objective, that is, independent of the human observer and the process of knowledge. By contrast, in the new paradigm, the observer, the process of knowledge and the phenomenon observed must be thought of as a system. In other words, epistemology is to be explicitly considered in any rigorous description of natural phenomena.

4) *A Shift from Building to Network as Metaphor of Knowledge.* Whereas in the old paradigm the metaphor of knowledge was of foundational values, fundamental laws, hierarchy of theories, etc. upon which the "building' of knowledge was erected, in the new paradigm, the metaphor of knowledge is a network of descriptions in which there are no hierarchies or foundations. Hence, rather than ruing the crumbling of foundations during paradigm shifts, adjustments in conceptual frameworks are to

be considered normal and, indeed, necessary for the accumulation of knowledge.

5) *A Shift from Truth to Approximate Descriptions*. Whereas in the Cartesian-Newtonian paradigm it was believed that scientific knowledge could achieve absolute certainty, in the new paradigm all concepts, theories, and findings are considered provisional and asymptotic to "the truth" at best. Furthermore, science cannot provide a complete and definitive understanding of the world because the experience of it can never be fully accounted for in scientific terms. (Capra and Steindl-Rast: xi-xv)

As can be seen from the above, the Ecosophical Paradigm offers a radically new perspective of reality than the Cartesian-Newtonian. It emphasizes wholes rather than parts, processes rather than structure, approximate descriptions rather than objective truth, complementarity rather than compartmentality, holism rather than reductionism, lateral thinking rather than linear thinking, whole-brain synthesis rather than left-brain analysis. But, and this is the crux of the issue at hand, the new paradigm can also be the basis of a new ethic based on harmony and respect for our home and all with whom we share it. It recognizes human responsibility in protecting, preserving and studying nature for ever-greater appreciation and benefit and, this being so, the Ecosophical perspective can provide the framework for a new, more responsible value-system for our time. In Sperry's words,

> Consider, for example, as a tentative starting maxim for determining right and wrong, to be accepted axiomatically without proof, something like the following:
>
>> "The grand design of nature perceived broadly in four dimensions to include the forces that move the universe and created man, with special focus on evolution in our own biosphere, is something intrinsically good that it is right to preserve and enhance, and wrong to destroy or degrade."
>
> With this start, defined strictly in terms that are scientifically sound, an extensive and coherent value-belief system can be constructed. Other axioms and propositions may be added as long as they are consistent. (Sperry: 1985, 22)

If we could commit ourselves to such a value-system, it would mark us, at long last, as a mature species, one that has put away its infantile obsession with itself and replaced it with a sense of empathy and moderation. We would do things differently, not because we are impelled to do so, but because we will deem it right. We would cease our paving over and grasping for things we don't need just as we outgrew our childhood fancies. We would wake up to the beauty and value of a simpler life and live more abundantly in doing so.

As improbable as all this seems, it is not impossible. Similar transformations have come about before, and whenever they did, the world was forever changed. ("Never underestimate the power of an ideal," Sperry reminds us). Nothing less than a species-wide conversion is in need today if we are to deal with the predicaments before us.

### The Educator's Most Important Role

Unfortunately, it takes time for paradigm-shifting ideas to take root in the consciousness of the general public. ("Really deep concepts seem to take about 50 years to sink into the collective conscience of the thinking community.")[94] The problem is that, as the world scientists remind us, we do not have much time at our disposal. We must learn and incorporate these new ideas soon, before disaster befalls us. We cannot wait in the hope of deliverance — we must help bring it about.

This is where education can play a crucial role. Its ubiquitous and well-established institutions are eminently placed for the efficient large-scale dissemination that this new doctrine requires. It is the key if we want to bring about the global mind change as quickly as possible, for, as Nelson Mandela has rightly said, there is no more powerful weapon for change in the world than education.

The foot soldiers of the schooling enterprise are educators. At their best, they personify those whom Matthew Arnold referred to as "the great men [and women] of culture,"

> Those who have ... a passion for diffusing, for making prevail, for carrying from one end of society to the other, the best knowledge, the best ideas of their time. (*Culture and Anarchy*, 1869)

Educators are the transmitters of all that has been achieved and has been worthwhile throughout the long, difficult history of humanity. Every teacher has been a link in what Mortimer J. Adler has called the Great Conversation between generations, imparting knowledge, challenging minds, shaping values, seeding future possibility. Every teacher, whether consciously aware of it or not, has had a critical role in the evolution of culture, which lateral and regressive steps notwithstanding, has generally advanced.

Today, educators continue in this honourable task, but their work will assume special importance. The shift from the Cartesian-Newtonian worldview to the Ecosophical Paradigm will be largely their responsibility to impart. They will be at the forefront in the war of values between the old world-view and the new; they will be the ones who will convey in spirit and in truth the new doctrine.

If they are effective in instilling its moral and if it is internalized in time, humanity may yet escape disaster, and civilization may be guided towards new levels of achievement and dignity. If, on the other hand, they will prove to be inadequate to the challenge, the future will be bleak. Much is at stake, and generations to come will judge them on how well they will have discharged this duty. But since it is up to the collective will to assign them the task quickly, we are all ultimately responsible for what may or may not come to pass.

Given the mandate, teachers will continue to teach phonics, mathematics, history, etc. as they always have, but they will integrate into their teaching the new narrative for our time. In so doing, and with a little luck, there will be a transformation of minds and values soon enough that it may alter our apparent destiny. This, as far as I'm concerned, is the single most important role that educators presently have. It is certainly more vital than teaching kids primarily how to make money to feed the insatiable beast of conspicuous consumption. We are facing an existential threat. Could anything matter more than coming to terms with it?

The mind change wanted will not happen overnight. But if the western nations were to take the lead in such an education initiative, it would be a huge improvement over the present situation for the simple reason that they are the major source of the current problems. China and India, which together constitute 36% of the world's population, could soon follow the example because their cultural traditions already have the inklings of the Ecosophical Paradigm mind-set. And we could be on our way. *It can be done if we so will it.* At the very least it is worth a try.

Of course, there will continue to be geopolitical differences, especially as resources become scarcer and the troubles begin to bite, but surely on this matter of ultimate concern it behooves us to act in concert. It is the only rational option if we are to prevail over what is to come.

Science has presented us the facts. They are before us and they are compelling. But as experience has shown again and again, facts by themselves do not necessarily lead to change in behaviour. Teachers from Aesop to Jesus, Harriet Beecher Stowe to Richard Rorty have reminded us that people recast their values not from moralizing, preaching or dialectics, but by empathy: identification with another through the faculty of the imagination. It is brought about, in the words of the naturalist E.O. Wilson, by striking "the inner mystic chord of emotion."

The ability to identify with others seems to be hardwired in us and in other species. (Brain scientists and cognitive neuroscientists have discovered mirror neurons, known as empathy neurons, that enable human beings to feel and experience another's situation as keenly as if it were one's own). One of the most effectual means of inducing this state of empathy and openness to change is *story*.

# PART VI

# THE KOSMOS STORY

## *The Consolation of Meaning*

Never in the history of humanity has there been more information available[95] or greater power over nature. Yet, because of the lack of a unifying, edifying myth,[ix] our lives are characterized by alienation and purposelessness. Towards the end of his life, Carl Jung reflected that most of the patients who went to see him were not so much mentally ill as in search for meaning.

This sense of meaninglessness has long been recognized as a dismaying condition of modern existence, and efforts in education have been made to address it, but with little success.

> The most common solution in cultural terms has been to reinstate past forms of humanistic studies in a core curriculum, a

---

[ix] I am using the term not in the sense of an invented tale, but as historians of religion Mircea Eliade and Joseph Campell advanced it, that is, as a way to experience the awe and nature of the universe, and the means for finding meaning and guidance in life.

curriculum which includes philosophy, ethics, history, literature, religious studies, and perhaps some form of general science — all of these in a critical rather than a commitment context. Yet somehow these programs do not seem to take. A cultural canon does not emerge that could do for our world what the religious and cultural orientations of earlier religious cultures did for the societies of those centuries. The program does not activate the human energies that are needed for a vital human mode of being. There is an inability to bring together the scientific secular world with the religious believing world or with the humanist cultural world. Each of these feels impelled to go its own way. Consequently all three remain trivialized. No unifying paradigm emerges. Effective education does not take place. No larger context is established in which the [school][96] can envisage itself or its educational mission. (Berry: 97-98)

What, then, can we do to remedy the situation?

At such crisis moments we need to return to the story of the universe. The entire [school] project can be seen as that of enabling the student to understand the immense story of the universe and the role of the student in creating the next phase of the story. Discovering this story has been the high privilege and central meaning of the modern scientific venture. (Berry: 97-98)

What is this story? It is the story of what American philosopher Ken Wilber refers to as the Kosmos,[97] all that exists, the sum total of all expressions of a creative universe as it unfolds through the various domains of matter, life and mind. It is, quite simply, the greatest epic ever known.

### *In The Beginning*

"Tell me a story," says the child, and the storyteller begins. In an instant, the world of common reality is left behind, and a new reality — more captivating, more intense, more real — catches up the listener on the wings of imagination.

We never as long as we live, stop saying, "Tell me a story." Our hunger is never satisfied; the more we read, the more we want to read; and the richer the feast, the hungrier we grow. For the master of creative imagination evokes the creativity in all of us, makes us all shareholders in the treasure that literature brings to life. The story — in prose or poetry, in art or music — is the magic of every man's life. By comparison, the most staggering achievements of science and industry and statesmanship seem to some people bodiless and cold. (Hutchins and Adler: 35)

Stories are powerful because they convey not only factual information, but emotional content. It is in the context of stories, as Joseph Campbell, Bateson, Berry and others have pointed out, that experience is organized, meaning imparted, knowledge made personal, psychology remolded — in short, that *life is made sense of*. And because we respond so well to stories, it is natural that the greatest tale ever known, as we shall see, that of nothing less than the history of the universe, should be taught as a story. It is in this that educators will assume the aforementioned crucial role: to tell the story, our story, the story of all there is, for the entire universe plays a part in this amazing adventure.

We are still in the process of transcribing the chronicle, but what we already know of it fills one with awe. Anyone who attends to it cannot help but be more reverent, more considerate of all that lives. It is the responsibility of educators, especially, to tell and retell the story until it reverberates inside the soul of every child and adult. It may literally be the first true doctrine of salvation.

The brief summary that follows is of the most rigorous and current of the countless theories about the history of the universe and our place in it that have been proposed over the ages. The details are far from precise[98] and the picture is not nearly complete.[99] But in outline, as the present generally accepted version of the tale has it, some $13.7 \pm 0.13$ billion years ago,[100] an incomprehensibly dense "virtual" particle billions of times smaller than a proton popped into existence with such energy that its expansion continues to this day and is expected to continue for at least another fifteen billion years.

At the moment of this so-called "Big Bang," space-time and all the matter-energy the universe will ever hold were spontaneously (that word again) created. After a relatively modest initial expansion (whereby the underlying thermal uniformity we presently observe throughout space as $2.725^\circ$ K was fixed) there was a sudden burst of accelerated inflation that within the first trillionth trillionth of a hundred billionth of a second ($10^{-35}$ seconds) had rendered the universe about the size of a grapefruit. Within the first second the expanding fireball had swelled to a sphere a thousand times the size of our solar system and had cooled from $10^{20}$ degrees K to $10^{10}$ degrees. It continued to expand at a somewhat slower rate, and eventually cooled enough for subatomic particle-waves (quarks, electrons, muons, neutrinos, and photons)[101] to precipitate. The four forces (strong, weak, electromagnetic, and gravity) disengaged from each other, and quarks combined to form protons and neutrons.

As the universe continued to expand, the radiation waves throughout were stretched, further cooling it. Three hundred thousand years after the Big Bang, the universe was a mere $2700°$ C,[102] and electrons slowed down enough to be captured by protons and be drawn into orbit around them. This allowed photons, which had hitherto been buffeted about the opaque space-filling plasma by the free-roaming electrons, to radiate outward, and there was light.

Hydrogen, the simplest atom, formed, and in time, groups of these atoms were drawn together by the forces of gravity and electromagnetism. As the clumps grew in size, they became more massive, thereby attracting even more hydrogen atoms to themselves. When the aggregates were large enough, the atoms became so closely packed that their nuclei overcame the repulsive electromagnetic force between them and they fused according to the constraints of the strong force, which operates over infinitesimally small distances.

The fusion of two hydrogen atoms into a single helium atom releases energy. When enough hydrogen atoms in sufficiently massive clusters fused this way, stars ignited. Drawn together by dark matter, stars began to group together. These groups, in turn, clustered into variously-shaped galaxies, of which there are today an estimated 100 billion in the observable universe,[103] each containing on average 100 billion stars.

Stars of different masses evolve differently. Some burn faster, others slower. All, however, eventually run out of the hydrogen that fuels their fusion into helium. When they do, depending on their size, some turn into white dwarfs, some into black dwarfs, others into red giants, still others into black holes. Stars a few times heavier than the sun and bigger evolve into supernovae, the explosions of which scatter as interstellar dust the atoms of all the natural elements that were forged, both inside the stars through the process of fusion and those that were created during the terrific explosion itself.[104] Drawn together by the force of gravity, the atoms of this stardust clump together and eventually form planets.[105]

On the outer fringes of one of the tens of billions of spiral galaxies in the universe, our own Milky Way galaxy, spins a middle-sized star[106] around which revolve nine planets and countless asteroids and comets. On the third planet nearest the sun, between 3.5 and 4 billion years ago, after a billion years of heating and cooling, collisions with extraterrestrial debris, intense electromagnetic radiation, volcanic activity and ceaseless rain,[107] certain configurations of organic (i.e. carbon-based) molecules operating far from equilibrium, were able to combine through trial and error to form catalytic webs. Some arrangements were interlocked into relational loops which, through feedback processes, became hypercycles.

### The Emergence of Life

Successive instabilities drove these chemical systems into evermore complex spontaneous formations. Some became dissipative (energetically open but structurally closed) structures capable of self-replicating. Chance rearrangement, sorting and folding of pertinent molecules led to a kind of proto-metabolism, and there eventually emerged, perhaps in iron sulfide deposits in ocean floor volcanic vents, a qualitatively new holon, and with it the new property we call "life". Its chief characteristic is that it is *autopoietic*, that is, simultaneously a product and agent of self-organizing processes. It came about when protein globules and nucleic acids, complex molecules with the capacity to replicate, mysteriously joined to form protoplasm.

The first organisms were non-nucleated bacteria, which were little more than jellylike suspensions of organic particles enclosed in semi-permeable membranes. These so-called *prokaryotes* survived by absorbing nutrients directly from the environment. Eventually, variations in temperature, concentrations of minerals, quantity of sunlight, etc. led to greater and greater diversification. As their numbers grew, they developed different strategies for survival.

One of the first innovations was fermentation, a metabolic process whereby complex carbohydrates (such as starches, sugars, cellulose) and nitrogen absorbed directly from the air, were broken down to produce ATP (adenosine triphosphate), an energy-carrying molecule, which they used to power their activity. The byproducts of the fermentation process included ethanol and carbon-dioxide. Other microorganisms evolved which used these waste products to fuel their own metabolic processes.

Eventually, food began to run out, and a second innovation, considered by some to be one of the most critical ever made by the biosphere, enabled organisms (such as blue-green algae) to trap light energy and convert it into ATP directly. This, of course, is photosynthesis. A byproduct of this process is oxygen. As more and more organisms produced oxygen, the gas formed a cover around the planet, which included a layer of ozone ($O_3$) that filtered harmful ultra-violet radiation from the sunlight. The oxygen, furthermore, combined with free hydrogen in the atmosphere to form water vapour, thereby binding the hydrogen (which otherwise would have escaped the gravitational pull of the Earth) to the surface of the planet. So successful were these organisms that within a few hundred million years, the $O_2$ content in the atmosphere rose from 0.0001% to 21%, the present-day concentration. Uncombined oxygen, however, was toxic to most of the organisms then existing. It rendered, as well, the atmosphere dangerously combustible. Numerous species became extinct and the entire bacteria domain had to reorganize to survive.

It did so by developing a metabolic process that made use of the oxygen. What had been a poisonous, unstable waste product now became a rich source of nourishment for some organisms. These so-called aerobic bacteria proved to be much more efficient producers of ATP than the anaerobic bacteria — whereas fermentation produces only two molecules of ATP for every sugar molecule broken down, respiration of the same sugar molecule can generate as many as thirty-six ATP molecules. Respiration was such a spectacular success that aerobic bacteria diverged into numerous different forms within a relatively short period of time.

Living in relative proximity and closely resembling each other, the bacteria often exchanged bits of genetic material when they bumped into each other. This led to even faster evolution, thereby opening the way for entirely new species to evolve. In the 1.5 to 2 billion years during which the prokaryotes were the Earth's sole inhabitants, they developed all of life's essential chemical processes: fermentation, the creation of ATP, photosynthesis, respiration, the removal of nitrogen from the air, and the ability to exchange genetic information. Life had taken a firm hold of the planet and made it its own.

About 2.2 billion years ago, *eukaryotes* (nucleated cells) evolved. Their basic difference from bacteria is that, whereas in prokaryotes genetic material was dispersed throughout the cytoplasm, the genetic material of the eukaryotes was contained in a membrane-encased nucleus. This concentrated the DNA, the macromolecule responsible for heredity, allowing for much more effective transfer of genetic information between organisms: sex was invented. As well, eukaryotes contained organelles (mitochondria, chloroplasts, ribosomes, lysosomes, etc.) which performed specific functions within the cell. This new cellular complex was the result of symbiotic partnerships between what had been different species of prokaryotes. Over time, the complementary operations of symbiotically related organisms became so well coordinated that they merged to become integrated units.

### The Evolution of Animals and Plants

With time, multicellular organisms became more numerous. They aggregated wherever they found food or other favourable conditions. As the collectives grew and became more complex, certain organisms within them began to specialize, performing specific tasks for the benefit of all. Coordination of these functions and communication within and among the cellular communities became more and more efficient. Eventually they began to function as multicellular organisms. These were the first animals, appearing around 700 million years ago. The earliest plants, multicellular organisms with chloroplasts (organelles that convert sunlight into ATP) evolved 200 hundred million years later. The animal and plant holons evolved along parallel lines, but there were some organisms (such as the present-day euglena) that shared characteristics of both.

As first, the animal communities were little more than globular concentrations of cells. In some cellular organizations, the chemical channels by which the cells conveyed information with each other eventually developed into a nervous system and, in time, its controlling centre localized to become a proto-brain.

As well, in order to regulate their metabolism and maintain a consistent internal environment, organisms had to remove excess calcium out of the water that flowed through them. Again, in another marvelous example of turning virtue out of necessity, they converted this waste material into shells, skeletons, body armour, teeth. A new evolutionary advantage resulted from this, and animals were now more robust, diversified, and ready to exploit new habitats.

Although both animals and plants originated in watery environments, periods of drought often stranded individual organisms on land. Some survived, and species slowly began to establish a foothold there. The plants arrived first, around 400 million years ago, turning the planet green. Animals, particularly those that fed on the plants followed a few million years later.

With the colonization of land by animals and plants there appeared a new life form in the holarchy: fungi. They played the important role[108] of feeding on dead and decaying organisms, thereby recycling waste material back into the biotic web. The animals, plants and fungi evolved together and became so successful in diversifying and proliferating all over the planet, that they are now considered three of the five "kingdoms"[109] of life, the broadest category of living organisms.[110]

Adapting to land was an extraordinary feat and shows, once more, the inexhaustible creativity of nature. New organs for breathing, tougher skin coverings, stronger bones and muscles, altered metabolisms, different behaviours, etc., developed for survival advantage and new opportunity. Not all species were able to evolve adequately, however. Some went extinct, others returned to the water. A few made both water and land their habitat. These were the first amphibians.

After some more millions of years, a branch of amphibians evolved into reptiles, animals whose entire life cycle takes place on dry land. Meanwhile insects, the first airborne animals, appeared. Plants developed sturdier structures to withstand gravity and grew larger and leafier to capture more sunlight. Many co-evolved with animals, which would eat their fruit and scatter the indigestible seeds far and wide, thereby allowing for the appropriation of new habitats. Some grew flowers and evolved strategies of reproduction that made use of airborne vectors, wind and insects, to disperse. Texture, smell, colour, taste and song proliferated.

The reptiles were the dominant large animal on the planet for 200 million years until a giant meteorite (12 km across) slammed near what is today the village of Chicxulub in the Yucatán peninsula 67 million years ago, causing the Earth to ring like a gong and altering its spin. The impact, estimated to equal 10,000 times the explosive power of all the world's nuclear weapons combined, so churned up the lower atmosphere with water vapour, debris and dust that it altered the climate of the planet and greatly reduced the amount of sunlight reaching its surface. This led to a major tear in the biotic web and another mass extinction, the fifth and last before the present age.

A tiny shrew-like animal that had evolved out of the Therapsids, curious mammal-like reptiles from millions of years before, was able to survive the catastrophe because, being small, it required little food relative to the enormous dinosaurs that we associate with the era. Also, being warm-blooded, it was able to maintain constant body temperature during the ensuing cooler and then much warmer, more humid global climate, something the reptiles[x] could not do. Finally, rather than laying eggs, the female gave birth to live offspring and then took care of its litter, thereby ensuring its survival. With the decline of the dinosaurs, these and the other mammals that branched out of them became the dominant land animal, feeding at night, hiding from large predators during the day.

---

[x] Controversial paleontologist Robert Bakker, who also argues that dinosaurs were warm-blooded, believes that the Jurassic Era's extinctions were due to disease and not as a result of the meteorite strike.

## *Primates and Humans*

In the post-Chicxulub world, the evolutionary advantages of the mammals allowed them to propagate and develop into numerous species within a relatively short period of time. Small early primates, known as prosimians ("pre-monkeys"), evolved in the tropics around 63 million years ago. Unlike other mammals their size, including some by now formidable predators, prosimians were not anatomically robust and, hence, were constantly threatened by enemies. They made up for this by developing greater agility, intelligence, and cooperation.

The prosimians were mostly insect eaters or vegetarian. Because they were primarily arboreal, they developed binocular vision, prehensile limbs and the bigger brains that went with them. When food was scarce, they would climb down from the trees to forage the ground.

About 35 million years ago, they bifurcated into the lower primates (lemurs, lorises, tarsiers) and pre-anthropods. The latter, in turn, split 20 million years ago into monkeys (with tails) and apes (without). The apes were mostly ground creatures. Ever on the lookout for predators, they would assume upright posture for short periods of time.

Around 5 million years ago, the apes branched into orangutans, gorillas, gibbons, chimpanzees and baboons. A million years later, a chimpanzee species in the African savannas evolved into an upright-walking ape, *Australopithecus* ("southern ape") *africanus*.

An important difference between human beings and other primates is that human infants have a much longer childhood. Whereas the young of most other mammals (marsupials are one exception) leave the womb more or less ready for the outside world, human infants are helpless at birth and remain so for years. In short, compared with other animals, human babies seem to be born prematurely.

This suggests that a relatively hairless species of primates with a much less fixed brain structure (depending for its survival on learning and experience rather than instinct) evolved from apes prone to premature pregnancies. Consequently, the need for supportive, multi-generational communities. Females selected males who would provide for them while they nursed and protected their infants. Moreover, as the *oestrus* cycle in females waned, they became sexually receptive throughout the year and, thus, promiscuity in both sexes diminished. Single-partner sexual relationships became the norm, and the family unit was reinforced.

The first true human type was *Homo habilis* ("skillful human") who branched out of *Australopithecus* between 2 and 2.6 million years ago. This species of hominids was the first to make[111] (primarily stone) tools. About 1.6 million years ago, another humanoid, *Homo erectus* ("upright human") branched out of the *Australopithecine* trunk. The members of this new species not only made more sophisticated tools than any before, but they also hunted in groups, built shelters and, importantly, learned to control fire.

*Homo erectus* was also a wanderer, migrating from Africa into Asia, Indonesia, and Europe. Successive ice ages advanced and receded and, in order to survive, communities of *Homo erectus* learned to band together as extended families and cooperated as never before.

Between 400,000 and 250,000 years ago, *Homo sapiens* ("wise human") bifurcated out of the *Homo erectus* branch. This is the species to which modern humans belong, the only surviving member of over a dozen hominid species that came and went, at times coexisting, often competing for limited resources, occasionally cross-breeding.[112]

*Homo sapiens sapiens*, our species proper, appeared around 200,000 years ago in sub-Sahara Africa. These people manufactured fairly sophisticated tools, using wood and bone as well as a variety of stone, lived in cohesive social groups, used fire to cook meat and keep warm, stitched together clothes from hides, and began to display signs of a spiritual life, as attested by ritualized burial grounds. Intimations of self-consciousness begin to stir.

Precisely how, in the evolution of hominids, the neuronal networks achieved the degree of complexity that allowed for the emergence of "self" is controversial. It seems as though the system of neurons that gave rise to the interior life of humans was first assisted by reflex-type responses to stimuli (fight or flight), then by images, and finally by an observer capable of organizing and deliberating on those images. Once this reflective ability came to mind, motivations, emotions, preferences, etc. came to be controlled by a self-conscious agent with concerns beyond mere survival. Culture began to take root.

Around 50,000 years ago, *Homo sapiens sapiens* began to use language, unquestionably the most important factor in the development of culture.[113] Language set humanity apart from all other animals and became a new factor driving its evolution. The phylogenetic roots of language are not yet clear, but because it plays such an important role in making us what we are, it should be dealt with in some detail.

## *The Development of Language*

New experimental methods and instruments (including cameras capable of tracking eye-movement and, hence, attention span) show that babies as young as just a few days can distinguish the difference between one, two, and three. This suggests that the capacity for numerical sense is innate,[114] and so it may be for language, as the linguist Noam Chomsky has argued.[115]

We know, as well, that distinguishing among various sensations ("differentiating type" in the jargon) is something all living organisms do simply by virtue of the fact that they must respond appropriately to different stimuli in order to stay alive. (Recall Bateson's and the Santiago Theory's identification of life with mind). Humans are unique in that we are adept at distinguishing large number of types from the endless stream of perceptions our senses acquaint us with moment by moment. According to linguist-mathematician Keith Devlin, this ability as well as our erect posture and the subsequent velar closure of the oral cavity (which permitted the articulation of consonants and, therefore, a much wider variety of sounds) led to a proto-language of object-type direct verbal pointings ("antelope there", "storm clouds ahead").

This type of real-time stimulus-response thinking developed more advanced frameworks for identifying patterns in (abstracting) and communicating about (symbolizing) an ever-increasing number of simple forms of information exchange. Driven by curiosity, a desire to exchange information, an effort to be understood better and, importantly for Devlin, a passion for gossip, eventually shifted the abstracting and symbolizing frameworks to "off-line" thinking: the brain learned to bypass the need for direct stimulus in order to produce meaningful utterances. That is, it developed the ability to generate its own stimuli and operate on them without need of input from the external world. Ideas and feelings became as effective impulses to articulation as sense perceptions had been.

But evidence suggests that off-line thinking preceded language for hundreds of thousands of years. Why did it take so long for language to arise? As Devlin explains,

[Y]ou need a lot of types before you can think off-line. In addition, you need a mechanism for combining them in a way that corresponds to the *structure* of real-world situations. What kinds of combinations are you likely to need? Well, what kinds of connections between things do we encounter in real-world situations that give those situations their structure? Here are some: things stand in various kinds of relations to other things, things act on other things, things combine to act on other things, things precede other things in time, actions cause other actions, actions prohibit other actions. But these are precisely the things that syntax gives you with its subjects, verbs, objects, and clausal structure. In other words, the combinatory machinery necessary to initiate and maintain off-line thinking is nothing other than syntax." (Devlin: 244)

What this means is that before true language could emerge, the brain had to be able to represent a sufficiently rich world structure consisting not only of a great variety of different types, but of *relationships* among them. The schematic model of the world that the brain simulated is what gave rise to syntax, and it is the mapping of syntactic structure onto proto-language that obtained language. This proved to be a lengthy process because, as primatological evidence indicates, proto-language and language proper are different in kind. To get from one to the other required a jump across a chasm. What made it possible?

According to Devlin, the capacity for language was likely preceded by a reorganization of neuronal connections, particularly in the left frontal lobe. This was likely due not to any major mutation in the brain. Rather, the central circuitry required for syntax may have already been in place for some other function and was co-opted for the purpose of combining words into grammatical sentences. The adoption of an organ for some purpose other than it was intentionally used for is a common phenomenon in evolution and has been termed by Stephen Jay Gould "exaptation" (in contrast to "adaptation").

> In some cases it seems that organs develop to serve one purpose and when they have reached a certain form in the evolutionary process, become available for different purposes, at which point the process of natural selection may refine them further for these purposes. (Chomsky, 1988: 167)

The cerebral reorganization that led to syntax may have occurred when the brain exapted the circuits for generating sounds and used them to structure grammatical sentences. The evidence supporting this view is that both these abilities are located in Broca's area of the brain's left frontal lobe. Damage to this region leads to the impairment of both speech production and the ability to handle grammar even in the abstract, that is, using non-verbal symbols.

Regardless whether Devlin's or some other version can best explain the mechanics behind the phylogenetic acquisition of language, there is substantial anthropological evidence indicating that the language faculty evolved and came to be selected because it provided survival advantage. It soon became an essential characteristic of the species along and co-evolving with abstract thinking and symbolic representation. In due course, it proved to be the cornerstone of culture, the primary means by which complex ideas took shape, knowledge imparted, learning accumulated over time. The exchange of information became much more efficient and collective wisdom accelerated. Humans began to act in unison like never before, and before long, they began to understand, preempt, and eventually control nature.

New tools and techniques were devised, animals tamed, agriculture invented, settlements established and soon after, civilization, commerce, writing, mathematics, philosophy, science, technology. A new world, Popper's

World 3, the realm of ideas and culture, emerged and flourished. And here we are, in the blink of an eye in terms of the biotic history of the world, the dominant force on the planet — yet dependent on a biosphere that fills, surrounds and supports us, and whose intrinsic value only now we are beginning to appreciate.

This, in brief, has been the story of our heritage. It begins neither with the Sumerian civilization, nor with *Australopithecus africanus*, nor with the first animal that flopped on shore, but from the very first moment of creation. It is the story of the Kosmos, and this story in all its grandeur and mystery tells how we came to be who we are: the spectators and actors in a great epic, both a marvel of creation and merely one other species on a planet that suppurates life. It reminds us how improbable, indeed miraculous, our ascent has been. It is a story that can teach us to be more conscientious in our treatment of the world and all the creatures with which we share it and depend on.[116] It is a story that reveals to us our identity as children of the universe, glorified star dust, kin to all yet uniquely endowed with power and responsibility. We, as mythologist Joseph Campbell was fond of saying, are as though fishing for minnows and not realizing that we are riding a whale. The Kosmos Story can awaken us all to this.

By telling and retelling the tale, starting with the very young we can reshape, as Thomas Berry has argued, the structure of our consciousness. We become bigger, more reverent, heroic. It ennobles us. It satisfies our need to feel part of something timeless, over and above our petty cares and our fretting selves. It feeds the hunger that organized religion can no longer assuage. Indeed, it can awaken in us the "cosmic religious feeling" of mystics, saints and those dedicated to the pursuit of truth that Einstein spoke of, by arousing in us a sense of wonder and gratitude.

> While this account is scientific, it is also mythic as a coherent presentation of the universe against backgrounds far beyond anything that rational intelligence can properly understand. (Berry: 98)

In the light of the new narrative, life becomes imbued with a new sense of what it means to be human. We learn to see the world differently, and as William James pointed out, how we interpret the world makes all the difference as to how we live in it.

# PART VII

## THE MYTHICAL DIMENSION OF LIFE

### *The Universe as Our Home and Parent*

In the early part of this century, Bertrand Russell wrote,

> That man is the product of causes which had no provision of the end they were achieving; that his origins, his growth, his hopes and fears, his loves and beliefs are but the outcome of accidental collocations of atoms; that no fire, no heroism, no intensity of thought and feeling, can preserve an individual life beyond the grave; that all the labours of the ages, all the devotion ... all the noonday brightness of human genius are destined to extinction[117]... (as quoted in Augros and Stanciu: xiv)

From the perspective of the Ecosophical Paradigm, this view is unduly pessimistic. There is now greater trust in the wisdom of the universe. And while the new story says little regarding "life beyond the grave," Hamlet's "There are more things in heaven and earth, Horatio, than are dreamt of in your philosophy" now seems an apt answer to questions of ultimate destiny.

Notwithstanding all that we have learned, existence remains an enigma within a mystery. As geneticist J. B. S. Haldane famously said about the universe, not only is it queerer than we suppose, but queerer than we *can* suppose. But there is less angst. We now have a deeper appreciation of the cosmos and our place in it: we are part of a timeless ever-unfolding drama far greater than anything we can imagine. It is an optimistic view, perhaps the most comforting since Copernicus pried man from his position at the heart of creation. Although our situation is not necessarily central, it is privileged simply by virtue of the fact that, improbable as it seems, we exist in a universe that, at least in our nook of space-time, is friendly to life and in large part is *comprehensible*.

We are not incidental castaways in an alien, absurd universe subject to blind physical laws to which we are deterministically bound etc., etc. Rather, we are autonomous agents in a cosmos that is suffused with intelligence, creativity and staggering complexity to which we have insight and, on occasion, can understand in exquisite detail. It is a universe whose order, amazingly, our minds are somehow tuned to accord with. ("The most incomprehensible thing about the world is that it is comprehensible," Einstein famously said). In a word, it is a cosmos that would be known, and in this there is comfort. As eminent theoretical physicist Paul Davies has said,

> The fact that … we human beings are privy to the hidden principles on which the universe runs, seems to me to be a fact of profound significance. It would have been easy for biological evolution to produce organisms that were intelligent but nevertheless unable to crack the mathematical code in which the laws of nature are encrypted. … The fact that we can come to know the laws suggests to me that our existence in the universe as conscious organisms is not merely an incidental quirk of fate, but is fundamental to the workings of nature. In other words, our own existence is intimately related to the existence of the universe, with its particular laws and structures. That is not to say that *homo sapiens* as such is preordained, only that the emergence of mind from matter somewhere is written into the laws of the universe in a basic way. … Whatever meaningless Darwinian accidents may have contributed to *homo sapiens*' characteristics, the existence of our rational minds can be no incidental triviality.[118]

We are, furthermore, in intimate communion with all there is, was, and ever will be. The universe is our home and parent. With prodigious patience and meticulous attention to detail, it has laboured to give birth to us,[119] and through us, as Thomas Berry has put it, it has come to reflect upon its own amazing being. The universe may ultimately be indifferent to our hopes, our fears, our joys, but we are that part of the universe, however small, which is *not* indifferent, and herein lies the beginning of meaning. This, in essence, is what the Ecosophical Paradigm proffers.

If we would but recognize and keep reminding ourselves that we are an expression of the awesome power of the universe, we would treat this world and each other, with greater respect and much less grief. We could be cured of our grasping, self-engrossed mania that is at the root of so many personal and social ills. We would appreciate life and nature at the deeper, more reverent level that Martin Buber, Thoreau, Arne Naess, E.O. Wilson, and saints and mystics have spoken of. We would recapture the joy and wonder of childhood in the simple experience of being alive. We would dedicate ourselves to what is truly worthwhile, rather than squandering our precious time on worthless pursuits. In the words of cosmologist-philosopher Brian Swimme,

[I]f we would attend to the task of creating work and forms of culture and forms of society that would enable us to remember the magnificence of existence in ordinary life, then all of the craziness and obsession in consumerism would just drop away like a disease that's run its course. And we would ease out of that and into this form of consciousness that just takes deep delight in ordinary life … [I]f you think about the amount of energy we're putting into consumerism — you work really, really hard and without thinking you end up surrounded by a whole bunch of stuff you never really wanted! If we find ways to ease out of that compulsion, then we have all this energy for creating a life that's simple, that's compassionate, that's profoundly spiritual. Moment by moment. And then instead of all that energy being trained on destruction, it would all be drawn back into the true work of the human — which is to be where beauty is deeply felt. That to me is the essence of the Epic of Evolution. (Swimme: 1997, 5)

To be fully human, E.O. Wilson has said, requires that a person view life in a heroic dimension. This is what the Kosmos Story, or what Swimme refers to as the Epic of Evolution, can provide. It is the loom upon which the numerous threads of our lives, which are now tangled and frayed by countless pulls, can be woven into a fabric of meaning and wisdom. It is a story sufficiently legitimate and authoritative to render moral guidance, purpose, a sense of continuity, a vision for the future. It is, in short, the kind of metanarrative that can provide meaning and heal us of the personal and collective malaise that is driving us to spiritual and, perhaps, literal suicide.

Let us be clear about this: life is resilient. It will survive and continue to thrive on this planet regardless what we humans do or not do. There is not the remotest chance that we will destroy all life on earth. The dangers at hand affect and will affect but a fraction, albeit a significant one, of the species on this planet. But that includes our own. We too will be subject to the same ecological upheavals forecasted, and this will almost certainly lead to destruction, conflict, significantly lower living standards, widespread misery. It will be the start of a decline the slope and extent of which can only be guessed. It may even lead to our extinction long before our due time.

### *Tat Tvam Asi!*

So, it is as much to save ourselves from ourselves that we have to hear the story and live by its moral. The story must be told early, it must be told often. And at each subsequent telling, suitable additions in detail and precision must be made to reinforce and render more tangible its wonder and meaning. The brief account given in this book is but the roughest sketch of its magnificence. It has merely provided the warp threads upon which teachers and anyone who tells the story can weave and embroider as much as their knowledge and craft permit. (Competence in multiple disciplines would be an asset here).

The story, moreover, should be promulgated and augmented by every technological means we have at our disposal, from audiovisual media to social networks. An all-out effort must be made so that the message is spread as far and wide as quickly as possible. Let us consider it as the basic doctrine of a new faith in humanity for the salvation of the world, and for our collective soul.

We are the dominant species on the planet, but there is no moral justification for treating as objects at our disposal, as we do, the other creatures with which we share our world. They too have needs and desires like us: food, shelter, companionship, freedom of movement, avoidance of pain. To deny them such without any sense of consideration exemplifies the worst form of prejudice and strikes at the ethical centre of what we idealize our species to be.

But suppose that we are willing to assume a narrower, more brutal definition of ourselves and continue to allow our interests to trample upon those of the rest of the natural world, what then?

As previously mentioned, regardless how harmful our actions prove to be for countless species, however diminished the biotic community will be because of us, life as a whole is too resilient to be totally destroyed by humanity. It will continue, and organisms, beetles and flies, for example, which together constitute nearly half of all animal species, or ants and termites, which constitute a third of the terrestrial animal biomass, will respond and adapt appropriately to the changes we are inflicting on the environment and thrive on. We, however, do not have the thousands, let alone the hundreds of thousands, of years required for fundamental genotypic modification.

If climate change or any of the numerous toxins and now gene-modified organisms[120] we are introducing into the environment turn out to be particularly baneful, we could be wiped out as a species, just like all our hominid cousins were. We number seven billion but we are not hermetically sealed from the rest of the biosphere. We depend on countless other organisms to keep alive. We could neither feed ourselves let alone digest our food without their help. Should the finely-balanced processes that the web of life has developed over millions of years be sufficiently disturbed (and not much is needed for such to happen because feedback loops and non-linear equations, as we have seen, can greatly amplify even slight changes) it could lead to irreversible calamity. Consider:

It is well known that microorganisms are super-specialized, often feeding on just one particular food, which could be deadly for any other organism. If, say, one of the hundreds of the new chemicals we introduce into the world every year should prove to be toxic for a species of such creatures, the microorganisms which had been its food could proliferate so rapidly that it could lead to the collapse of an entire ecosystem, or even the overwhelming of the entire biosphere. Preposterous, one may think. Far from it. A bacterium that doubles every ½ hour (a common growth rate among prokaryotes) would result in one week's time into a colony numbering $2^{336}$, which is $10^{21}$ times the total estimated number of atoms in the universe![121] (Sagan: 219)

The probability that any similar outcome could literally happen is minimal, thanks to the many checks and balances in nature. Yet the example is apt because it demonstrates quite dramatically how quickly ecological collapse can occur in the non-linear processes that operate within ecosystems. The same non-linear dynamics operate in climate change. The point is that the problems facing us are gathering momentum, and they are converging. We must mobilize to blunt the impact. We must change abruptly, deliberately, collectively. We must learn to see the world we inhabit in a responsible, respectful way. This is a heroic and inspiring stance that will provide us with meaning and spiritual sustenance as we work to undo the damage we have wrought.

The Kosmos Story, encompassing all of life, the Earth, the entire universe can provide the metanarrative of the first truly global culture on this planet. It can serve as the energizing myth of our age, a mantra for transcendence. It awakens us to the marvelous fact that we are no after-thought, an incidental byproduct of atoms colliding at random, but a significant, integral component of the Earth community, which includes the atmosphere, hydrosphere, geosphere, biosphere and mindsphere, within a magnificently complex universe that is unfolding in a common destiny, a singular adventure. *Tat Tvam Asi!*[122] "That art Thou! It is your story that is being told. You are a part of it all!" This is this grand and wondrous message that the Kosmos Story relates. It is a privilege fraught with responsibility for us, in this critical time in the history of the biosphere, to have come to know it. May we be worthy of its wisdom.

## *Conclusion*

The more we learn, the more we cannot help but be amazed by the twists and turns of the past 13.7 billion years that have brought us to this point in space and time. The evolution of humanity is the result of numberless contingencies and was by no means guaranteed. Nor is our survival from here. Genetically, we are practically indistinguishable from other primates,[xi] but thanks to our larger frontal lobe, opposable thumbs and a capacity for language, we, in a remarkably brief period of time, have acquired an unprecedented degree of power over nature.

---

[xi] We share 98.6% of our genome sequence with the bonobo chimp.

But we are now in the midst of an ever-accelerating rush towards evermore consumption, and have lost a sense of proportion. This has done and continues to do great harm to our fellow creatures and our home's natural processes, so much so that reactive momentum is gathering apace. Approximately 5000 generations of humans have come and gone since we first ventured out of Africa 100,000 years ago. It will be up to the present generation to determine whether humanity continues in its ascent towards the stars or begins a descent to eventual oblivion. We must come to terms with this momentous fact, and we will have to decide one way or the other. The choice is stark: one path leads to almost certain disaster, the other not. It is clear which of the two we would choose, but the question is will we do so in time?

Failing to act as we should in this twilight hour of our gifted age, our species may be living through the last stretch of reasonably good days for a long spell. We could be brought to our knees never to rise again. All the low-hanging fruit have been plucked. A series of extreme climatic events can be so destructive, can so disrupt our now largely urban infrastructure, that supplying the needs of any sizable fraction of seven billion people would be an enormous logistical challenge.[xii] Sustenance would have to be eked out of marginal lands and more polluted waters. The raw materials required for the products we can no longer do without would have to come from sparse, more inaccessible sources. It would be an era of great privation and interminable strife, a Hobbesian nightmare from which there is no waking. Certainly, not a future we would wish upon our children. History is a chronicle of surprises. This time around what awaits us is largely foretold. The surprise is how poorly we seem to be preparing for it.

And *even if* the scientists, environmentalists and economists turn out to be mistaken Cassandras, it is prudent that we err on the side of caution and give heed to the prevailing scientific consensus. There is no more credible alternative. By any measure, our present course is immensely destructive, unsustainable, and morally inexcusable. I think that we all suspect this. Yet we carry blithely on lemming-like towards the precipice.

No-one will be able to cure us of this collective madness that has seemingly overtaken us. We must do it on our own, individual by individual. But we will change our

---

[xii] Note how feebly even rich and powerful America was able to respond to 2005 Hurricane Katrina, or Japan to the recent tsunami.

ways only if we are compelled or have overwhelming reason to do so. Popular culture is largely fashioned by commercial interests, and their agenda resists tampering with the *status quo*. The political and social institutions are also largely beholden to exuberant consumerism.

Our best hope is an alternative way of viewing and treating the natural world than we currently have. In my opinion, there is none better than the one offered by the Kosmos Story. It is the means by which the insights of the Ecosophical Paradigm can most cogently be transmitted, internalized, clear eyes, open minds, embrace the world and help mend it. As such, it should be a fundamental component of education, a *leitmotif* running through it all, explicitly written in the Ministry guidelines and embedded in the curriculum of every grade and every level. It should be a core subject at teachers' colleges. And it should be promoted to other school boards across the world, which should then encourage one another to join in the great enterprise of its effective dissemination.

It is in sharing the message with single-minded purpose and diligence that we, hopefully, can help humanity continue to participate positively in the great cosmic venture that has engendered us. A quixotic endeavour? Perhaps.

But what other alternative is there? Surely we can do more than merely wait and rage against the coming night. We have to try *something* to hold it back as best we can. We stand to lose so much otherwise — not only the awesome beauty and luxuriance found everywhere in nature, but also Archimedes, Leonardo, Shakespeare, Mozart, Mother Teresa, and all else that has and would have been achieved ever since that first human left his handprint in some remote cave. Let us not snuff out the torch he lit, which has come to illumine so brightly a little corner of this vast and wondrous universe. Let us continue with some measure of grace and dignity our uniquely uncharted journey that, notwithstanding all the mayhem and grief we have caused along the way, has manifested sublime achievement and boundless possibility. There is no more important task, in my opinion, than to share the story and live by its spirit. If we do so in time, we may yet keep humankind from becoming a glowworm that was consumed by its own light.

- 163 -

## *Endnotes*

---

[1] "1992 World Scientists' Warning to Humanity"    (www.ucsusa.org/about/1992-world-scientists.html)

[2] In a 2010 Angus Reid poll asking whether global warming is a fact and caused by human activities, only 58% of Canadians, 41% of Americans, and 38% of Britons agreed. And these are among the most educated and well-informed people on Earth.    (angusreid.com/polls/38822/views_on_global_warming_vary_in_three_countries)

[3] Earth Policy Institute. March 2, 2004.    (earth-policy.org/index.php?/plan_b_updates/2004/update35)

[4] "Is Biodiversity at Risk?"    (www.biodiversitybc.org /EN/main/ why/110.html)

[5] "UN says case for saving species more powerful than climate change." (guardian.co.uk: May 21, 2010)
   A recent study by the International Program on the State of the Ocean asserted that "dead zones" due to lack of oxygen, warming and acidification are affecting the world's seas "faster than the worst case scenarios that were predicted just a few years ago." It cited a single bleaching incident in 1998 that killed one-sixth of the world's tropical coral reefs, adding that unless immediate action is taken to protect them, all coral reefs could be gone by 2050. "We now face losing marine species and entire marine ecosystems ... within a single generation," concluded its authors.    ("World's Oceans in Dire State." *The Toronto Star*. June 20, 2011)

[6] In the 3.5 billion years or so that life has existed on this planet, the geological record shows at least five mass extinctions: about 440 million years ago, when 85% of marine animal species were wiped out;  367 million years ago, when a majority of fish species and up to 70% of marine invertebrates vanished;  245 million years ago, when 95% of all life perished;  208 million years ago when another substantial fraction of sea and land species was wiped out;  and 65 million years ago when 75% of all species, including the dinosaurs became extinct.
(Earth Policy Institute. "The Sixth Great Extinction: A Status Report." www.earth-policy.com/index.php?/plan_b_updates/2004/update35)

---

[7] Species extinction rates have increased quite substantially in the last few decades. Between 8000 BCE and AD 1600, it is estimated that one species disappeared every 1000 years. Between 1600 and 1900, one species became extinct every four years. In the early years of the 21st Century it is estimated that one species is becoming extinct every 30 minutes. Presently, 19,000 species are in danger of extinction, up from 11,000 in the year 2000.

According to the IUCN (International Union for Conservation of Nature) Red List for 2010, 33% of all species of the major groups of organisms are currently "threatened" (i.e. critically endangered, endangered, or vulnerable). Specifically, 20% of all vertebrates (21% of mammals, 12% of birds, 21% of reptiles, 30% of amphibians, 21% of fishes); 30% of all invertebrates (insects, mollusks, arachnids, etc.); 68% of all plants (resulting in the loss of an inestimable amount of genetic information that could lead to medical, chemical or bioengineering breakthroughs); 50% of fungi and protists (essential for the robustness of all ecosystems).                    (www.iucnredlist.org/documents/ summarystatistics/2010_4RL_Stats_Table_1)

What is worth remembering in pondering these statistics is that once a species is extinct, not only is it gone forever, but its absence creates an imbalance in its former habitat. In the oft-cited analogy of rivets in an airplane wing, the loss of a species has relatively little effect on the ecosystem as a whole. But if several species disappear, the system would essentially collapse as does an airplane wing on losing too many rivets.

[8] "The first decade of the twenty-first century was the hottest since recordkeeping began in 1880. With an average global temperature of 14.52 Celsius (58.14 Fahrenheit), this decade was 0.20 degrees Celsius (0.36 degrees Fahrenheit) warmer than any the previous decade. The year 2005 was the hottest on record, while 2007 and 2009 tied for second hottest. In fact, 9 of the 10 warmest years on record occurred in the past decade."    ("Past Decade the Hottest on Record."www.earth-policy.org/index.php?/indicators/C51/global_temperature_2010)

The 2007 Fourth Assessment Report of the UN Intergovernmental Panel on Climate Change (IPCC), produced by over 600 authors from 40 countries and reviewed by 620 experts and governments concluded that "Warming of the climate system is unequivocal" and that "most of the observed increase in global average temperatures since the mid-20th century is very likely (i.e. 90%) due to the observed increase in

anthropogenic greenhouse gas concentrations." The report also stated that average global temperature is expected to rise 2–3° C by 2050 provided emissions peak soon. Inaction would lead to a 6° C increase by 2100. The effects of the forecasted global warming include a 40% extinction rate for animal species, and 200 million people forced from native lands due to rising sea levels and droughts.
(www.climateandfuel.com/pages/news.htm)

[9] In 2011, the international Arctic Monitoring and Assessment Program reported that the melting of the Arctic and Greenland ice caps are projected to raise sea levels by 35 to 63 inches (0.9 m to 1.6 m) by the end of the century. This is a sharp increase from the 7 to 23 inches estimated in 2007 by the U.N.'s scientific panel on climate change. The reason for the new figures is accelerated warming: the past six years have been the warmest period ever recorded in the Arctic, and the warmest in 2000 years, according to ice core studies. (These have also shown that the thickness of the ice caps has shrunk by half since 1979).

Such a sea-level rise would threaten the coasts of low-lying countries, including Bangladesh and Indonesia, two of the most populous nations in the world, islands everywhere, and cities from Shanghai to London.
("Sea levels could rise 2–3 feet more, Arctic experts say." May 3, 2011.www.msnbc.msn.com/id/42878011/ns/us_news-environment)

[10] Exceptionally hot weather was blamed for a record 9.5 million acres that burned in the U.S. in 2006.

Warmer air also holds more water vapour, resulting in bigger storms. In the last quarter century there have been 164 Atlantic hurricanes, a 34% increase over previous 25-year periods. The 2005 Hurricane Katrina, a category 5 (catastrophic) storm, was economically the most destructive in U.S. history. Hurricanes Mitch (1998) and Wilma (2005) were also category 5, the highest on the scale of hurricane strength. Although less powerful, hurricanes Charley and Ivan (2004) and Rita (2005) caused billions of dollars worth of damage as well.

There has been, furthermore, a marked increase in the number of tornadoes. In April 2011, in the worst tornado breakout in the region in nearly a century, a number of twisters wracked havoc through southern U.S., killing over 300 people and causing billions of dollars worth of damage. A few weeks later, a tornado through Missouri killed over 120, making it the deadliest single tornado in 60 years.

[11] (www.bbc.co.uk/news/world-latin-america-13449792)

While tropical rainforests cover only 2% of the planet's surface, they are home to two-thirds of all living species on Earth, supporting up to 1000 species per square kilometer. For example, it is estimated that the Andean mountain range and the Amazon jungle are home to more than half the world's species of flora and fauna. And there are more fish species in the Amazon river system than in the entire Atlantic Ocean.

[12] Fossil fuel emissions hit record highs in 2010, topping 30 gigatons, about 5% more than the previous record set in 2008, before the global recession.                                                    (www.usatoday.com/tech/news/2011-05-31-carbon-emissions-hit-record_n.html)

[13] There is a loose-knit but influential group of high-level scientists and assorted "experts" who claim that global warming and other ecological threats have been invented for financial or ideological reasons, including higher taxes and greater government control. A number of fine books (*Merchants of Doubt, Deceit and Denial, Climate Cover-Up*, among others) deal with the phenomenon of "manufactured ignorance" whereby these so-called "deniers," politically conservative scientists with strong connections in big business and politics, use their scientific credentials to discredit well-established scientific knowledge, spread false information, sow doubt and confusion among the public, and attack the reputation of opponents. This stance is amplified by right-wing media.

As has been the case in the past with similar refusal to accept overwhelming evidence regarding the dangers of cigarette smoking, the use of pesticides, the cause of acid rain, chlorofluorocarbons' effect on the ozone layer, etc. so it is now with the current ecological problems. Organized, ardently committed and well-funded by powerful lobby groups, these skeptics have disproportionate influence over the general population because they allow the masses to justify a preferred though harmful course of action on the basis of lack of scientific consensus when, in fact, there largely is. Bruce Babbit, US Secretary of the Interior under the Clinton administration called attention to one such example when he said in 1998:

> It's an unhappy fact that the oil companies and the coal companies in the United States have joined in a conspiracy to hire pseudo scientists to deny the facts …
> the energy companies need to be called to account

because what they are doing is un-American in the most basic sense. They are compromising our future by misrepresenting the facts by suborning scientists onto their payrolls and attempting to mislead the American people. (wikipedia.org/wiki/Global_warming_conspiracy_theory)

Although useful as gadflies on the body politic, these professional skeptics must ultimately be dismissed along with their conspiracy theories and misrepresentation of facts. There have always been those who criticize some policy while continuing to reap its benefits. Nor can any argument or evidence convince the willfully blind (n.b. black emancipation, women's rights, labour laws). The world has progressed in spite of such people. So it must now, for the stakes are much too high.

One of the more respectable environmental skeptics, Bjørn Lomborg, is right to say that humanity is resourceful and has a habit of technological innovation that has resolved numerous problems which at first seemed catastrophic. But to argue, as he does, that the present problems, too, will come to be solved in due time without the need of radical change is Pollyannaish. There are numerous examples throughout history when seemingly robust civilizations (Maya, Easter Islands, Moche, etc.) suddenly collapsed, very likely from ecological causes. Nor does his assertion that prosperous countries may be sufficiently resilient to the natural disasters due to global warming reassure poorer countries, which also happen to be, quite often, the most populated. And, of course, none of this is any comfort to the species being wiped out by the day just so that we need not cut back on carbon emissions "too soon."

[14] Some of the remedies that, in my opinion, would begin to improve the present situation include, in no particular order: a moratorium on deep-sea trawling; replacing coal with natural gas or nuclear energy while we develop alternate energy sources, especially solar; extending the adoption of genetically modified crops to improve yield while minimizing fertilizer use and further land clearing; more public transportation, green roofs, energy-efficient lighting, better insulation, more extensive recycling; less meat; a graduated carbon tax, proceeds from which are to go towards offsetting losses in declining industries (eg. coal mining), encouraging sustainability (eg. protecting rain forests, wetlands, marine habitats, etc); and, of course, continued research.

[15] "The idea of man's conquest of nature is inseparable from sexual imagery, as feminist scholars have made very clear. ... Using metaphors derived from contemporary techniques of interrogation and torture of witches, [Francis Bacon] proclaimed that nature 'exhibits herself more clearly under the trials and vexations of art [mechanical devices] than from when left to herself.' In the inquisition of truth, nature's secret 'holes and corners' were to be entered and penetrated. Nature was to be 'bound into service' and made a 'slave' and 'put in constraint.' She would be 'dissected,' and by the mechanical arts and the hand of man, she could be 'forced out of her natural state and squeezed and molded,' so that 'human knowledge and human power meet as one.' He advised the new class of natural philosophers to follow the model of miners and smiths in their interrogation and alteration of nature, 'the one searching into the bowels of nature, the other shaping nature as on an anvil.' And he wrote of the new science as a 'masculine birth' that will issue in a 'blessed race of Heroes and Supermen.'" (Sheldrake: 1992, 43)

[16] "Another year, another billion." *The Economist.* The World in 2011

[17] Global Footprint Network  (www.newcommunityproject.org/consuming_appetites.shtml)

[18] ibid

[19] ibid

[20] Harman and Sahtouris: 240

[21] Earth Policy Institute. March 2, 2004.  (earth-policy.org/index.php?/plan_b_updates/2004/update35)

[22] "Welcome to a Zero-Sum World." *The Economist.* The World in 2011

[23] As a philosophical position, this notion, anticipated as early as 5th century BCE by Protagoras of Abdera ("Man is the measure of all things") and in more recent times by Nietzsche's "Perspectivism," is now a cornerstone of postmodern, feminist, and other historicism-based epistemologies that claim that all knowledge is "situated" within

implicitly circumscribed contexts. In other words, there is no *absolute* standard of reference of any kind.

[24] Einstein's theories produced a revolution not only in science, but in the general understanding of life and in the arts. The initial reaction to the rejection of an absolute standard for measuring space and time was despair, moral relativism and even solipsism. In literature, the realist novel, which had been a natural offshoot of Cartesian-Newtonian physics, was replaced by the psychological novel, which depicted states of mind rather than an "objective" reality. Two examplars of this literary genre are James Joyce and Marcel Proust. Cubism and Dadaism are examples from the visual arts. The atonal compositions of Schoenberg, Webern, and Berg are examples in music.

[25] This is likely due to virtual particle fluctuations in the underlying vacuum field. See endnote 99.

[26] The formulation of this fact $(\Delta x \Delta p \geq h)$ is known as Heisenberg's Principle of Uncertainty, posited in 1926 by the German physicist Werner Heisenberg. It means that it is impossible to measure both the position and the momentum of a particle simultaneously with more than a strictly limited degree of precision. Specifically, the product of the uncertainties in position and momentum must always be greater than or equal to Planck's constant $h$ (i.e. $6.62606896 \times 10^{-34}$ J-s).

One of the many implications of the Uncertainty Principle is that there is no longer a one-to-one relationship between initial and final conditions. In other words, two identical particles with identical initial states may end up having any one of many different final states.

Planck's constant $h$ is the number that links the energy of a particle with the frequency of that particle's associated wave. It, furthermore, marks the boundary of what we can know: below this value's equivalent in dimension and time, the laws of physics no longer apply.

[27] The fundamental role of mind in structuring experienced phenomena was first given rigorous philosophical treatment by Immanuel Kant (1724-1804) in his *Critique of Pure Reason*. He proposed twelve "categories of understanding" by which the mind structures order upon the bewildering flood of sensory experience. The categories are quantity (unity, plurality, totality), quality (reality, negation, limitation), relation

(substance, cause, community), and modality (possibility, existence, necessity). It is through the mental template of these categories, and what he called "pure forms of sensible intuition," i.e. space and time, also mental constructs, that we make sense of the world — indeed, that we can say we experience it at all.

Kant's notion that space and time are wholly mental constructs has now been refuted: space-time is an actual, if not fundamental, physical entity, the context or background for all particle interactions and their concomitant forces, and it is possible to detect its dynamics by measuring how a given mass of matter curves it or, equivalently, how it causes that matter to move. The observation that proved this fact, the first (indirect) detection of gravitational waves, led to the 1993 Nobel Prize in physics.

An entirely new and radical approach to the metaphysical role of mind has been suggested by Information Theory, which proposes that reality is composed not of matter, energy, forces, etc. of standard physics, but of binary choices ("bits") of information. For other intriguing theories regarding the nature of reality, see Deutsch and Paster.

[28] "Complementarity" is the term used to refer to two mutually exclusive ways of understanding something, each of which, being valid within its own particular context. Neils Bohr enunciated the complementarity principle in 1927 as an essential feature of Quantum Mechanics. An experiment measuring one aspect of a system of atomic or smaller dimension (such as particularity) will preclude the possibility of learning anything about a "complementary" aspect (such as waveness) of the same system. Yet both aspects are necessary and needed for a more complete understanding of the quantum event.

[29] The proof is not considered conclusive because of still-unresolved problems with "local hidden variables". Whether these are technical problems or problems in principle is still an open question.

[30] Bohm, as quoted in Wilber: 1982, 129

[31] Bohm's work and the discovery that memory is diffuse throughout the brain led the neuroscientist Karl Pribram in the 1970s to conclude that the brain is itself a hologram perceiving, interpreting, and participating in a holographic universe. Our brains "mathematically construct concrete

reality by interpreting frequencies from another dimension, a realm of meaningful, patterned primary reality that transcends time and space." (Wilber: 1982, 5)

For some, Pribram's account would allow for a more precise mechanism by which consciousness "brings forth a world" or "creates" a reality than the standard Quantum Mechanics explanation that the mind somehow "collapses the quantum wave function" to bring the world into being. (see de Quincey)

[32] This notion is a contemporary reformulation of the *Great Chain of Being* (or what Ken Wilber refers to as the *Great Nest of Being*), which, according to the theologian Houston Smith, is a tenet common to all of the world's great wisdom traditions. (Wilber: 1998, 6)

[33] *Dualism* is a theory that maintains that substances are either ideal or material, neither category being reducible to the other. Dualism is distinguished from *monism*, which asserts that all that exists is fundamentally one and only one substance; and *pluralism*, which claims that the world consists of many different kinds of existents, which in their uniqueness cannot be reduced to either one (as in monism) or two (as in dualism) fundamental substances.

[34] In contrast to his teacher Plato, Aristotle believed that neither form nor matter can exist apart from each other but are separable only by abstraction. Form is immanent in matter and the means by which the essential nature of all things, existing only potentially in matter, is expressed and made real in the world. Aristotle called the process of this self-actualization towards some goal of development *entelechy,* that is, "self-completion".

Entelechies have been referred to by various authors as vital factors, life forces, formative impulses, *élan vital*, souls, *chi*, etc.

[35] The historian Morris Berman has called the machine model of organisms "mechanomorphism," the projection of mechanical characteristics onto all of nature. But mechanics, Berman observed, is an invention of just one species, human beings, and so mechanomorphism is an even worse form of anthropomorphism than usual, for we are attributing a single activity of a singular species to the entire natural world.

---

[36] In his 1976 book *The Selfish Gene*, Dawkins also introduced the concept of "meme," a unit of cultural ideas, symbols, or practices conveyed from mind to mind through speech, ritual or other form of communication. According to Dawkins, memes transmit ideas or belief information in a manner analogous to the way genes transmit genetic information. As well, they self-replicate, mutate, adapt or perish according to selective pressures similar to the way genes evolve through biological evolution.

[37] This idea is reminiscent of Carl Jung's Collective Unconscious, an inherited component of the psyche to which every human being has access. The Collective Unconscious is a kind of cumulative memory of the species that stores, organizes and channels the personal experiences of every individual in a common mode, giving rise, thereby, to universal themes (archetypes) in dreams, myths, and patterns of behaviour.

[38] In a series of experiments begun at Harvard University and continued over a 50-year period, rats trained to escape from a maze learned more easily with each passing generation. But even more surprisingly, rats of the same genetic lineage in other parts of the world learned to escape from comparable mazes faster than the original rats did. Similar types of patterned behaviour or accelerated learning have been observed in groups of primates, birds (eg. milk-drinking bluetits), and isolated human communities, which often have similar near-simultaneous intellectual insights. (www.sheldrake.org/Articles&Papers/ papers/morphic/morphic_intro.html.)

[39] "Mind, Memory, and Archetype Morphic Resonance and the Collective Unconscious." Rupert Sheldrake, *Psychological Perspectives*. Spring 1987. (www.sheldrake.org/Articles&Papers/ papers/morphic/morphic1_paper.html)

[40] For a non-biological example, it has long been observed by synthetic chemists that when a new substance is produced, it is generally quite difficult to crystallize. Once done however, subsequent crystallization, occurs more easily wherever it is attempted. The conventional explanation has been that fragments of previous crystals spread through the atmosphere as microscopic dust particles to "seed" the new crystals or, even more improbably, that crystal fragments were carried from

laboratory to laboratory on the beards of migratory chemists (sic!) Sheldrakes' explanation is that a totally new substance at first has no morphic field to conform with and, therefore has no guidelines, as it were, to follow. But in time, as a morphic field emerges and gets more firmly established, the substance finds it easier to attune to and harmonize with. For other examples, see Sheldrake: 1981.

[41] The phenomenon whereby two dispersed genetically identical seeds in close proximity germinate at different times during a given season so that the resulting plants do not compete with each other for available resources.

[42] The mathematical concept behind the feedback loop is the "recursive function" whereby the independent variables the function operates on are results of the same function's operations of the preceding step. For example the function $x_{n+2} = x_{n+1} + x_n$ where $n \in \mathbf{N}$, $x_1 = 1$ and $x_2 = 1$ gives rise to the Fibonacci Sequence $\{1, 1, 2, 3, 5, 8, 13, ...\}$ because $x_3 = x_2 + x_1$ generates the number 2, then $x_4 = x_3 + x_2$ generates the number 3, and $x_5 = x_4 + x_3$ generates the number 5, and $x_6 = x_5 + x_4$ generates the number 8, and so on. The function, in other words, is self-sustaining; once initial values are assigned, it generates new values by feeding on its own output, in principle, forever.

The recursive function, essentially an abstract formulation of self-referentiality, has proven to be one of the most fruitful concepts of the $20^{th}$ century, crucial not only in cybernetics, but also in Gödel's theorems, computer science, fractal geometry, Chaos Theory, etc.

[43] By "pattern" what is meant is not an ostensible design or arrangement of things, such as the pattern of stripes on a zebra or that trees have roots and leaves. Rather, what is meant by pattern is an *abstract* condition of regularity. The circularity of feedback loops in machines and living systems is an abstraction and not a literally observable pattern of roundness. The other fundamental pattern of organization inherent in living systems is a network of internal and external *relationships*.

Perhaps a better word would have been Bateson's "metapattern," a pattern of patterns, a pattern that connects all living creatures, but not in terms of quantities and shapes, forms, and relations. (Bateson: 8-11)

[44] More precisely, it came from one of two different schools of thought in cybernetics, the one associated with Norbert Wiener. The other school, associated with John von Neumann, considered living organisms as data-processing machines. This latter perspective was widely influential until Gregory Bateson (who, along with his wife, the anthropologist Margaret Mead, had taken part in many of the discussions among the early cyberneticists) revived and developed Wiener's view as an alternative to the organism-as-computer model.

[45] In simple terms, linear equations are equations in which the highest power or degree of any of the variables is the number 1, such as $z = 2w + 3x - y$. In nonlinear equations the degree of any one or more variables is other than 1. For example, $z = 2w^{5/3} + 3x^2 - y$. The important thing here is that in linear equations, a change in any of the independent variables ($w$, $x$, $y$) obtains a proportional change in the dependent variable $z$. In non-linear equations, on the other hand, a small change in the independent variable(s) can lead to large differences in the dependent variable. This result is formalized in Chaos Theory as The Law of Sensitive Dependence on Initial Conditions.

[46] These are also known as chemical clocks, whereby chemical reactions produce sudden periodic oscillations from one state to another, analogous to toggling a switch back and forth.

[47] This process is considered by many to offer a much more persuasive explanation of the mechanism driving evolution than the standard Darwinian appeal to gradual change through random mutation and natural selection, which are deemed too slow to have generated the enormous variety of life we witness in so young a planet.

It would also explain the bursts of evolutionary activity after long periods of stability that we see in the fossil record. Paleontologists Stephen Jay Gould and Niles Eldredge proposed the Theory of Punctuated Equilibria to account for such evidence. The theory asserts that most sexually reproducing species experience little evolutionary change over most of their history. This stasis ends with rare and sudden branching called *cladogenesis* whereby a species splits into two distinct species, rather than gradually transforming into another, as the standard theory of evolution has held.

[48] There is also the possibility that life began elsewhere and was brought to Earth via some asteroid, or spread through "directed panspermia" by alien intelligence, as the biologist Francis Crick proposed. The strongest evidence for an extraterrestrial genesis of life comes from a recently-published report on a fragment of a meteorite that exploded in early 2000 over Lake Tagish in the Yukon. Scientists found "a surprisingly large variation" of amino acids and monocarboxylic acids, both essential for life. This suggests to some that the chemical building blocks that gave rise to life on Earth may have "aged to perfection" in asteroids. (www.msnbc.msn.com/id/43344767/ns/technology_and_science)

Nonetheless, the general consensus among scientists is that unless there is overwhelming evidence to the contrary, the working hypothesis is that life on Earth originated on this planet.

[49] The phenomenon of function preceding structure is, according to some, strong evidence of goal-oriented evolution. In other words, evolution is the unfoldment of solutions to problems, and not merely a sifting out of useful traits out of purely random mutations, as standard evolution theory has it. (see Harman and Sahtouris: 94-95, 149).

> [I]f bacteria are placed in a medium with nutrient molecules too large to pass through the pores of its membrane, they may mutate to make the membrane more permeable. Bacteria grown in a salty medium become better able to survive and reproduce in seawater ... *Escherichia coli* unable to digest lactose become able, through "purposeful mutation," to metabolize it when they had to ... [W]hen E. coli in a solution of salicin was given little nourishment, it underwent two otherwise rare mutations together at a rate thousands of times higher than in normal growth, making itself able to use the salicin. [Biologist Barry Hall] suggested "that cells have some means of recognizing what would be an advantageous mutation and increasing the chance that it occurs." (Harman and Sahtouris: 45)

[50] Varela, as quoted in Capra: 1996, 98

---

[51] This essential difference was recognized as early as Kant, who also proposed the notion of "self-organization" as a definition of organisms.

[52] Computer simulations of living systems (such as *Life*, a cellular automata model) and so-called "virtual pets" appear to display remarkably life-like behaviour. Of course, their existence does not depend on the interrelationship of their physical components as is the case with natural living systems, but on the flow of electricity through microprocessors according to strictly defined algorithms that is then translated visually onto a monitor screen. For a fascinating account of *Life* and other examples of surprising emergent properties of similar mechanical generative systems, see Holland.

The surprisingly complex results from the iteration of relatively spare algorithms led physicist Stephen Wolfram in his 2002 *A New Kind of Science* to conclude that the universe is fundamentally digital (rather than analog) and that the processes and laws of nature can best be described by means of simple computational modules rather than the daunting differential equations of traditional mathematics.

[53] Maturana, as quoted in Capra: 1996, 267

[54] Humans, furthermore, share a language and a conceptual framework through which, as the Russian psychologist Lev Vygotsky pointed out, we bring forth a world *together*. In other words, ours is also a socially-constructed reality, having both species-wide and culturally-specific characteristics.

[55] "Evidently, the universe is characterized by an uneven distribution of causal and other types of linkage between its parts; that is, there are regions of dense linkage separated from each other by regions of less dense linkage. It may be that there are necessarily and inevitably processes which are responsive to the density of interconnection so that density is increased or sparsity is made more sparse. In such a case, the universe would necessarily present an appearance in which wholes would be bounded by the relative sparseness of their interconnection." (Bateson: 38)

[56] The view that the chief characteristics of nature are competition and survival of the fittest ("nature, red in tooth and claw") is now outdated.

Competition is only one aspect of nature's organization into holonic networks. As the British essayist William Hazlitt (1778-1830) pointed out, altruism is as natural and common among organisms as competition. Symbiosis, co-evolution, and cooperation in the pursuit of mutual self-interests are now recognized as common features of the natural world. "Life did not take over the globe by combat, but by networking." (Margulis, as quoted in Capra: 1996, 232)

For example, mitochondria and plastids (eg. chloroplasts), a cell's power-generating organelles, are, evidently, descendents of what were once-independent organisms that billions of years ago joined, for shelter and food, the now-containing cell. Evidence for this include: mitochondria and plastids are similar in size to prokaryotes; both mitochondria and plastids contain DNA different from that of the cell's nucleus; they are surrounded by two or more membranes, the innermost of which differ from the cell's membranes; that their ribosomes (sites of protein manufacture) are different than those of the containing cells.

It's also been proposed that the tail-like apparatus in many single cells is due to another primordial symbiotic union, this time with ancestors of *spirochete*, "cork-screw bacteria" that spiral in rapid motion.

There are numerous examples of cooperation and symbiosis in the macro-world as well (sea anenomes and hermit crabs, bees and orchids, African oxpeckers, cleaner fish). Indeed, according to Lynn Margulis's theory of "symbiogenesis," the creation of new forms of life through permanent symbiotic arrangements, are the principal means of evolution for all higher organisms, and a much more powerful evolutionary drive than competition. (Capra: 1996, 231)

[57] According to the Russian geologist Vladimir I. Vernadsky, organisms have created and incorporated 99.9% of all the kinds of organic molecules found in nature, almost all billions of years ago when bacteria were the only life form. These very same molecules have been used again and again by successive generations of organisms. "[E]volution … has been a matter of rearranging the same molecules and life processes into an endless variety of new creature patterns." (Harman and Sahtouris: 78)

[58] Recently, evolution has generated only species. No new phyla have appeared since the early Cambrian period (590 million years ago); no

new classes for at least 150 million years; no new orders for 60 million years. (Harman and Sahtouris: 99).

[59] Capra: 1996, 229

[60] Viruses, inhabiting a twilight world between the living and the nonliving are not autopoietic, and, hence, not alive, because they are incapable of independent existence. Consisting only of a protein coat and a core of nucleic acid (either DNA or RNA), they lack energy-producing protein-synthesizing capabilities, and can grow and reproduce only when they enter a living cell. On doing so, however, they suddenly spur into activity. They subvert the host's metabolism so that viral reproduction is favoured in their stead, and quickly change form in order to evade antibodies intent on neutralizing them.

[61] Grof: 7

[62] Sahtouris's notion was anticipated by Giordano Bruno (1548-1600) an excommunicated Dominican monk. Against Aristotle's view that matter is passive and inert, Bruno argued that matter is intrinsically active and self-transforming. In Bruno's cosmology, "consciousness" or "intelligence" is a necessary and intrinsic facet of nature. For him, mind is not an epiphenomenon of matter but is inherent in the very stuff of matter: it is mind-within-matter. Bruno was burned at the stake for these and other "heretical thoughts".

[63] en.wikiquote.org/wiki/Max_Planck

[64] That nature may be intelligent or even self-conscious does not necessarily imply that there is some overall design or pre-established goal. The future seems fully open because, insofar as Quantum Mechanics has shown, randomness is an integral feature of physical reality at its most fundamental. This plus the creative aspect of life makes the future fully contingent.

Of course, because of the non-locality nature of the cosmos, what may appear to us as a random event may have been determined by causes elsewhere in the universe. Since we can never know those causes, the event would remain for us purely random. (The concept of "cause," incidentally, is itself problematic, as David Hume pointed out.)

Even more intriguing, recent experiments appear to show that the occurrence of phenomena at a specific moment in time may have been determined by *future* events. As bizarre as this notion seems, "reverse causality" is gaining increasing support. Briefly, experiments have shown that measurements performed in the future can influence results that happen before those measurements were made, as though "ripples of the measurements carried out in the future could beat back to ... combine with effects from the past" to produce results in the present. If this turns out to be true, not only is the assumed one-directional "arrow of time" wrong, but it would also imply that the universe might have a "destiny" to which all is drawn.                    ("Back to the Future." *Discover*. April, 2010)

[65] Harman and Sahtouris: 257

[66] "Consciousness and the End of the War Between Science and Religion."                    (www.huffingtonpost.com/deepak-chopra/consciousness-and-the-end_b_620133.html)

[67] This view is much like the subjective idealism of British philosopher George Berkeley, who denied the existence of material substances. Things in the physical world, he argued, like tables and chairs, stars and starfish are only ideas in the minds of perceivers and, as such, cannot exist without being perceived.  For Berkeley, *Esse est percipi*, that is, "To be is to be perceived."

[68] What we have in these and similar views is essentially a secularized version of pantheism (eg. Spinoza's God *as* nature) or panentheism, (God *in* nature) has been substituted with Intelligence, Spirit, Life, etc.

[69] A recent experiment by Jeff Lundeen and Aephraim Steinberg of the University of Toronto suggests that an observer may not be needed after all for the Schröedinger equation of a quantum state to collapse.  They were able to probe reality at the subatomic level using so-called "weak measurements" without disturbing it and, hence, by Heisenberg's Uncertainty Principle, without "observing" it. They found that a definite state had obtained from the quantum process nonetheless. This adds weight to the "reality" side of Bells's Inequality Theorem and, therefore,

indirectly supports Alain Aspect's non-locality results. ("I'm Not Looking, Honest!" *The Economist*. March 9, 2011)

Aspect's results, incidentally, have been further supported by Yakir Aharonov and colleagues, who have shown that particles can be affected by electromagnetic fields even in regions far beyond those fields' expected reach.

[70] At the atomic level there is no definite boundary between I and the chair I am sitting on or the clothes I am wearing. Atoms from my body and those of my clothes and chair freely and continuously trade places, combine, break apart, disappear to reappear elsewhere. At the quantum scale, we and our immediate surroundings are part of and enveloped in a turbulent ocean of vibrating, spinning particles hurtling through electromagnetic fields that crisscross space-time. Our sensory receptors gather *qualia* from this helter-skelter cauldron of energy and mind shapes it into some coherent experience of entities and/or processes. As to how arbitrary parts of this indistinguishable continuum takes on the appearance of what we experience as distinct, separate wholes (such as chairs, clothes and humans) remains a mystery of the first order.

[71] For a superb treatise on this subject, see Horgan's *The End of Science*.

[72] The primacy of mind also features in the work of Descartes, Leibniz, Berkeley, Schöpenhauer, Bergson, and in some interpretations of Quantum Mechanics. And, of course, countless mystics and thinkers.

[73] The philosopher Edmund Husserl (1859-1938) was the first modern thinker to attempt a systematic study of the contents of the mind, but his approach was phenomenological, that is concerned with description of the phenomena that appear in consciousness and analysis of its structures, rather than providing scientific explanations.

[74] The bicameral brain is cross-wired so that the left hemisphere controls the right side of the body, and the right hemisphere controls the left.

[75] Even in so-called instances of Multiple Personality Disorders a person normally experiences herself (most cases of MPD are female) as a single subject even though she may display markedly disparate behaviours at different times.

---

[76] A striking consequence of this revolution, as has been previously mentioned, has been the displacement of physics as the exemplar *par excellence* for doing science. Life and mind, and not matter and forces, are now considered central phenomena, and these are understood to require methods and conceptual models quite different than those useful in the study of physics.

[77] The legal restrictions imposed on psychotropic drugs in the mid 1960s practically eliminated scientific studies of the effect of these substances on humans. Grof was later able to develop a technology ("Holotropic Breathwork") that induces similar experiences, and ensuing benefits, without the need of drugs.

[78] Although Penfield at first believed that memory was located in specific areas of the brain, subsequent experiments led him to change his opinion. Indeed, as the American neurophysiologist Karl Lashley showed, removing even half a rat's brain has practically no effect on the retention of its learning. This has been confirmed in so-called hydrocephalic ("water in the brain") geniuses who have merely a thin (1 mm) film of brain cells. Memory and mind seem to be everywhere and nowhere in the brain. Such conclusions suggest for many that the brain is more like a radio receiver than a memory storage system.

[79] Grof: 203

[80] In the mystical tradition this all-pervading ground is denoted variously as God, Light, Love, Divine Presence, etc. Deepak Chopra et al. refer to it as "pure awareness," the agent of creation and the source of every manifest quality in the universe. In Vedantic philosophy the sacred syllable "OM" both symbolizes and expresses this substratal reality.

[81] These constitute what Popper, in his *Objective Knowledge* (1972), referred to as "World 3" of his "three worlds of reality". "World 1" is the world of physical objects, the world of facts, the objective IT world. "World 2" is the world of consciousness, of mental processes and events, the subjective I world. "World 3" is the world produced by the human mind, the cultural WE world that includes works of art, ethical theories, social institutions, scientific theories, world views. Certain aspects of World 3 (e.g. mathematics) are "autonomous" in that they

operate according to their own dynamics, and human beings discover rather than create their truths. The relationships among the three worlds he describes as follows: "The three worlds are so related that the first two can interact, and that the last two can interact. Thus the first world and the third world cannot interact, save through the intervention of the second world, the world of subjective or personal experiences." (Popper, as quoted in Dykes: chapter 7: 2)

[82] Ken Wilber, who has sought to integrate the objective and the subjective, physics and metaphysics, science and mysticism, and eastern and western epistemologies, has devised a comprehensive map of reality, at least insofar as we experience it. His quadrifoliate model, the Four Faces of the Kosmos, demarcates the interior and exterior, the individual and collective dimensions. Each quadrant is then developed into a detailed description of that particular aspect. The result is what he calls an "integral vision" of the Kosmos.

[83] One positive outcome of the telecommunications and internet revolution of the last two decades is the extraordinary proliferation of NGOs and other grassroots movements throughout the world. Although they have been responsible for a number of small-scale initiatives and even some high-profile, if dubious, headlines (such as the Seattle and Prague demonstrations against globalism) the fact remains that national governments and multinational corporations still set the agenda and implement the policies that the majority of the world's people are subject to.

[84] From what has transpired since then, with America refusing to participate, let alone take the lead as would be expected of a superpower (responsible, by the way, along with China, for 40% of all greenhouse gases), in a number of international environmental initiatives, this statement seems less a rhetorical stance than U.S. government policy. But, lest we be tempted to bemoan those irresponsible Americans, a survey of 190 climate experts, released during the Cancún summit in 2010, ranked Canada 54 out of 57 nations in addressing climate change. As though to make its position perfectly clear, a few days after the Conservative majority victory in the Spring 2011 elections, the Harper government declared that Canada would not support an extension of the Kyoto Protocol on greenhouse emissions beyond 2012.

---

[85] The October 2006 700-page Stern Review on the Economics of Climate Change, the largest study ever undertaken on the economics of global warming, concluded that it would have cost at the time 1% per annum of the global gross domestic product (GDP) to avoid the worst effects of climate change. Late or inadequate response to the problem would result in a 20% drop in global GDP.

In March 2009, Professor Nicholas Stern, who headed the committee responsible for the report, increased to 2% GDP p.a. the estimated cost for stabilizing the $CO_2$ atmospheric concentration to 500–550 ppm because of faster-than-anticipated growth in emissions. He cautioned that, unless the opportunity was seized, even this target would slip away. Untimely action, on the other hand, was revised to precipitate a 30% drop in the GDP. ("Lord Stern on global warming: It's even worse than I thought." Michael McCarthy, *The Independent*: 13 March, 2009)

[86] The insidious nature of advertising and the enormous budgets devoted to it make it a formidable adversary in the war of values that, as we will see, educators especially will be obliged to lead.

While traditional values have generally waned in most cultures around the world, the cult of consumerism continues to grow unabated. Children are indoctrinated into it almost from birth, as evidenced by corporate logos on baby wear. Ronald McDonald, a cartoon dromedary, and the Nike swoosh are instantly recognized by children everywhere. (Indeed, according to former marketing executive and now consumer advocate Martin Lindstrom in his book *Brainwashed*, the average American three-year-old can recognize 100 corporate brands). Television programs directed to children are often nothing more than half-hour commercials for various toys. And now the internet, where children are enticed to visit web pages devoted to building enthusiasm for upcoming products.

[87] Global Footprint Network        (www.newcommunityproject.org/ consuming_appetites.shtml)

[88] A sense of meaning and place in the cosmos is prevalent among numerous traditional cultures throughout the world. In one example, there is a ritual among the Navajo people of Monument Valley in Arizona whereby an elder assumes the responsibility of welcoming the sun for the entire world community. The ceremony is by no means trivial. It involves a lengthy predawn run to a sacred place. Once the

---

preparatory purification has been completed, the sunrise is awaited. And then the welcoming ritual is performed with utmost diligence and solemnity, for if not done with impeccable care, the sun may not bestow its beneficence upon the earth's creatures.

The ceremony is one of many defining customs among the Navajo that have imbued the people with a sense of significance and obligation that touches every aspect of their lives.

[89] There is no generally accepted name for the new paradigm. The term "Ecological" is sometimes used, but because it refers to the environment, and because of conceptual baggage ("deep" ecology vs "shallow" ecology, social ecology, feminist ecology, etc.), the term would not be a suitable choice for what is considered a substantially cultural transformation.

"Ecumenical" has been proposed by some, including the theologian David Steindl-Rast, but it too carries conceptual baggage in the term's connection with the Christian church.

"Anthropocene," i.e. human-made, may accurately describe our time, but the concept is anthropocentric and, as we shall see, the new paradigm requires that we free ourselves of this bias.

Thomas Berry has proposed the term "Ecozoic" (Berry and Swimme: 1992, 241) in contrast to the Paleozoic, the Mesozoic, and the Cenozoic eras. But these represent geological epochs that range from 65 to 342 million years in duration. It would be absurd to think that the present paradigm could compare to those epochs in longevity.

[90] I conceived the term in the late-1990's and used it in a graduate paper in 2001, on which this book is based. I discovered at the time that the term was not original. A database search found it in connection with a religious-mystical organization. More recently, I have found that the term is being used to refer to

> [A] new variant of ecological philosophy based on an integrated ontological/epistomological approach. [It is] offered as a philosophical foundation for the political initiative to strive for ecological harmony with the more-than-human world. [It] attempt[s] to give the philosophical justification for the imperative of attaining such a relationship with our environment. [It] also serve[s] to disclose and foster a motivation for this effort by

providing a new vision of what human existence means. (www.thesighting.com/Ecosophy.html)
Happily, this use of the word coincides with my use of it.

[91] Kuhn defined paradigm as "a constellation of achievements — concepts, values, techniques, etc. — shared by a scientific community and used by that community to define legitimate problems and solutions." (Capra: 1996, 5)

[92] "Science advances one funeral at a time," Max Planck famously said.

[93] Another way to compare and contrast the old Cartesian-Newtonian paradigm (referred to as "Mechanism") with the Ecosophical Paradigm ("Holism") is outlined by John R. Battista (Wilber: 1982, 144).

| Parameters | Mechanism | Holism |
|---|---|---|
| Ontology | Dualistic | Monistic |
| Epistemology | Objective | Interactive |
| Methodology | Empirical | Analogical |
| Causality | Deterministic | Probabilistic |
| Analysis | Reductivistic | Structural |
| Dynamics | Entropic | Negentropic |

Still one other set of distinctions between old and new paradigms thinking is given by Marsha Sinetar.

| Traditional Mind | 21st - Century Mind |
|---|---|
| Egocentric frame of reference | Synergistic frame of reference |
| Split-perception: sees the part, focuses on detail, confused by paradox | Whole seeing: first sees whole then understands part, untroubled by paradox |
| Polarizes and separates; feels disconnected, alienated | Integrates, unifies; feels part of the whole |
| Fear-motivated | Love-motivated |
| Dualistic | Non-dualistic |

[94] Augros and Stanciu: xvi

---

[95] At the turn of the millennium, it was estimated that the amount of information available for consumption was doubling every six months. During the August 2010 Technomy conference in Lake Tahoe, Google CEO Eric Schmidt said that we now create every two days as much information as we did from the dawn of civilization up until 2003. (http://techcrunch.com/2010/08/04/schmidt-data/)

[96] I have substituted "school" for Berry's "college". I disagree with his (to me, curious) position that, "While this integral story is the proper context of the entire education process, it cannot be appreciated by students at elementary and high school levels in a reasoned, reflective manner." (Berry: 99)

If properly presented, the Kosmos Story can be appreciated by students of every age bracket, *especially* the youngest. It is at this age that they are at their most impressionable and particularly open to wonder. And as anyone who has dealt with children can attest, they can be remarkably perceptive *and* reflective, even as tots. What's more, as shown by the 200,000 young people who recently participated in the Green Your School Challenge in the U.S., and the iMatter-sponsored marches in 25 countries, young people are acutely aware of and deeply concerned about the environmental problems we face. As Alec Loorz, who founded the iMatter organization at age 13 has said, "Young people will be affected most by decisions that are made today and yet we can't vote, and we don't have money to compete with lobbyists … We do, however, have the moral authority and legal right to insist that our future be protected." This kind of mature yet idealistic attitude would be very receptive to the Story and should be cultivated.

("Fighting Climate Change Through Innovative Initiatives." May 2011, joanna.zelman@huffingtonpost.com)

[97] Wilber: 1996, 18-19

[98] One of the great difficulties in trying to make sense of reality is that a rigorous account of it can best be achieved through precise measurement, abstraction, and formalization — in other words, through mathematics. By means of it, science has been able to describe remarkably well *processes in space-time*. (Nobel physicist Richard Feynman likened the precision of Quantum Mechanics in describing physical processes to measuring the coastline of continental U.S. to

---

within a hair's breadth. Experimental results routinely agree with theoretical predictions to five decimal places and better).

Ordinary language, on the other hand, the language of narratives, describes phenomena in terms of *objects in space* and *events in time.* Interpretation of the well-defined terms of mathematics into ordinary language smudges their precision and accuracy. Unfortunately, whereas ordinary language is natural for us, mathematics requires long and difficult study, precluding the majority from understanding its results and implications at a sufficiently profound level.

Of course, as to why mathematics should be so adept at describing nature is another question. "If the structure of the math is deep, it will solve something in nature one way or the other … Every fundamental in math has ultimately had a meaning in the physical world." (Fields Medal winner Shing-Tung Yau). For an interesting treatment of the subject see Eugene P. Wigner's "The Unreasonable Effectiveness of Mathematics in the Natural Sciences." (Ferris: 526)

[99] For example, we still have no clue as to why there is so little anti-matter around, or even what constitutes 96% to 99% of the observable universe. It is reckoned that about 74% of it is made up of Dark Energy, an unknown force that for the last nine billion years has been pushing space outward and causing the universe's expansion to accelerate. The remaining 22% to 25% is made up of Dark Matter, which although invisible, must exist from the observation that galaxies hold together despite the fact that their mass is not sufficient for them to do so.

There is also the increasingly likely possibility that our universe is but one of countless slightly different and at times identical universes "parallel" to ours in an unimaginably vast multiverse. Or, if some of the interpretations of Quantum Mechanics are correct, that our universe is a bubble universe in an ever-expanding ever-bubbling megaverse. Or, as String Theory suggests, that it is but one universe among countless other universes hovering nearby, millimeters away but in other dimensions. In all cases, the number of universes is without end, perhaps approaching countable infinity, if not greater.

Furthermore, although the Big Bang is now generally accepted as fact among the scientific community, the state of the universe immediately after the Big Bang is still an open question. As for the Big Bang itself, a "fluctuation in the quantum vacuum" is about as precise a non-mathematical explanation as we have at this time. There is also evidence suggesting that the Big Bang may not have been a one-time event but is merely the latest of at least five similar creations.

Nor do we know much about the vacuum that permeates space-time. Far from being empty, it is an immensely dynamic field of virtual particles so energetically dense that a handful can boil all the oceans.

These are some of the mysteries of the macroworld. Insofar as the microworld is concerned, the so-called Standard Model has done a superb job of accounting for the particles and forces observed, but it is incomplete. We still do not know why certain particles are more massive than others. The Higgs boson, the theoretical particle that imparts mass to other particles as they burrow through the Higgs field, is yet to be found. So is the case with the graviton, the particle theoretically responsible for gravity. The Standard Model, moreover, is considered to be too mathematically inelegant to be the Theory of Everything (TOE).

One of the more promising candidates for the TOE is M-Theory, a composite of various versions of String Theory. It holds that at a fundamental level, matter consists not of zero-dimensional point-like particles, but of one-dimensional sinuous vibrating strings, which may be open or closed as in a loop, and "branes", their generalization to higher dimensions. Each vibrational mode of a string gives rise to a different point-like elementary particle (photon, quark, etc.) and its properties.

M-Theory not only describes a world that mathematically resembles ours, but also unifies gravity with the other three basic forces of physics, thereby at last reconciling General Relativity and Quantum Field Theory. It also explains *why* particles have the properties (such as mass, charge, and spin) they have. And it is consistent and elegant, two conditions considered indicative of the veracity of any physical law.

The rub is that M-theory posits 11 dimensions rather than the four dimensions (three for space and one for time) of our familiar world. Most problematic, however, is that many of its inferences cannot be experimentally verified. Because strings are so minute (for a rough estimate, according to physicist Michio Kaku, the ratio of the size of the universe to that of the earth equals the ratio of the size of the earth to that of a proton, equals the ratio of the size of a proton to that of a string), we would need a particle accelerator with a circumference roughly equal to Pluto's orbit around the sun to generate the kind of energy required to produce strings out of colliding particles. Still, for many, the theory is too elegant not to be correct, and work continues apace, despite the often intractable mathematics involved.

Although most physicists, including Einstein, believe that a TOE is achievable, not all agree. For Max Planck,

> Science cannot solve the ultimate mystery of nature. And that is because, in the last analysis, we ourselves are part of nature and therefore part of the mystery that we are trying to solve. (en.wikiquote.org/wiki/Max_Planck)

[100] We humans inhabit the so-called "middle-dimension" between the micro and macro worlds, which span 60 orders of magnitude. Our unaided eyes acquaint us with about 8 orders of magnitude, from the breadth of a hair to the distance from horizon to horizon. Units beyond these dimensions are so outside our ordinary range of experience that they have little meaning unless their scales are translated into terms familiar to us. For example, a billion, how big is it? If one had a billion dollars and she were to spend $1000 per day, it would take just about 2,740 years to spend it all.

[101] There are over two hundred types of subatomic particles known to exist, of which twenty-eight appear stable and indivisible. These particle-waves are in a perpetual frenzy of motion, transformation, popping in and out of existence, spinning, combining, splitting, absorbing and emitting energy. According to the bootstrap principle no particle-wave is more fundamental than any other because its existence is fully dependent on its *interrelationship* with all other wave-cles (another term for the same).

[102] For comparisons, the surface temperature of the sun is about 5600°C. And when the ITER nuclear fusion reactor, presently being built in southern France, is operative, it will be able to achieve temperatures in the range of 100 million to 150 million degrees Celsius.

[103] The observable universe is a sphere whose radius from us is equal to the speed of light times the present age of the universe. However, because during the first nanoseconds after the Big Bang the universe expanded at a rate much faster than the speed of light (while light is the fastest object that can travel *through* space, there is apparently no limit to how fast *space itself* can expand), its "outer edge" (estimated to be at least 300 billion light years across) is far beyond the scope of our observation, which, in principle, is 50 billion light years across. In short,

we can only observe what comes to us via light waves, and light waves beyond our cosmic horizon will never be able to reach us.

[104] Supernovae explosions, the most energetic in nature, are so bright that their light, equal to that of ten billion suns, outshines that of entire galaxies, sometimes for months on end.

[105] Once thought to be rare, it is now estimated that there are at least 50 billion planets in our galaxy. Some astronomers believe they may well outnumber stars. Most exoplanets thus far observed are large gaseous types similar to our own Jupiter. But numerous and varied as they are, it is highly probable that planets (or their moons) capable of supporting life also exist. There are, as well, billions of free-ranging planets that do not orbit any star. It is unlikely that these would harbour any life as we know it. ("There Are 50 Billion Planets in the Milky Way Galaxy." DigitalTrends.com. Jeffrey Van Camp, February 21, 2011)

"As we know it," incidentally, is an important qualification because life may have evolved differently elsewhere. For example, in 2010 scientists reported the discovery of a bacteria strain that uses arsenic instead of phosphorus in its genetic makeup. Up to that time, a golden rule in biochemistry was that life on Earth had to be composed of at least carbon, hydrogen, oxygen, nitrogen, sulphur and phosphorus. The finding is considered to be a major discovery because it shows not only that alternate life forms are possible and, therefore, more likely to exist on other planets, but that there may be a distinct "shadow biosphere" all around us in this world that we have either overlooked or wrongly assumed to be the standard kind.

[106] The Sun formed 4.5 billion years ago, relatively recently in the history of the universe. It has used up about half of its initial hydrogen supply. Because about 1% of it consists of elements heavier than hydrogen and helium (such as iron, carbon and neon) which could not have been produced in so young a star, it has to be a second or, even more likely, a third generation star made up in part of elements that were formed in previous stars, flung out during supernovae explosions and recycled in the sun's own fusion processes.

[107] Water, the most abundant liquid on this planet, is also one of the strangest molecules in nature. For example, rather than being positioned

symmetrically on opposite ends of an oxygen atom, the two hydrogen atoms are arranged closer together (at an angle of 104.5°) on one side of the $H_2O$ molecule, rendering it slightly polarized. In other words, every water molecule has a V-shaped structure whereby the two "feet" of the V are positively charged hydrogen atoms, and the tip of the V is the negatively charged oxygen atom. One of the many consequences of this configuration is that water is a near-universal dissolver, making it capable of breaking the chemical bonds of numerous molecules, which can then recombine with other molecules in different combinations. This was essential in the prebiotic combinatory processes of organic molecules that led to hypercycles and dissipative structures.

Another important consequence of its peculiar architecture is that water is one of two substances (the other is bismuth) that are less dense in solid state than in liquid state. During the numerous ice ages that occurred on Earth, the surface of small and large bodies of water froze over. Yet aquatic life below the surface continued undisturbed. Were ice denser than liquid water, it would have sunk and lakes and rivers would have frozen from the bottom up. This would have exposed the animals and plants living in the water to drastic environmental changes that would have been deadly to many species. Furthermore, as ice ages receded and temperatures rose, the sun's rays would not have been able to heat the ice at the bottom of large bodies of water, and most of it would have remained solidly frozen. With the advent of new ice ages, more ice would have piled onto the ice below, and the level of liquid water would have continued to shrink, forcing life into an ever-smaller domain. Again, this would have drastically reduced the number of species, and evolutionary creativity would have been greatly curtailed.

Perhaps strangest of all and most critical to life is the fact that the melting and boiling points of water are much higher than would be expected by comparison with structurally related compounds. For example, whereas the boiling point of $H_2O$ is 100°C, for $H_2S$ it is −60.3°C. What this means is that there should never have been any liquid water on the surface of the earth at all — it should all have been in a gaseous state. Were it not for these and other anomalous properties of water, life, as we know it, could not have emerged on this planet, let alone evolve to any great complexity.

[108] Fungi may be important in another quite different way. The remarkable affinity between the psilocin molecule (produced by the

psilocybin compound) and neuroreceptors in the human brain suggests to psychedelics researcher Terence McKenna that the 200+ species of mushrooms that produce psilocybin had a significant role in the evolution of the human brain (by far the most complex object in the universe yet discovered, consisting of $10^{11}$ neurons, engendering as many connections, via synapses, as there are stars in 15 thousand Milky Way galaxies), and hence, more complex mental content.

[109] According to the Carl Linnaeus (1707-1778) system of classification, the Five Kingdoms of Life consist of Monera (prokaryotes, i.e. non-nucleated single-cell organisms such as bacteria and blue-green algae); Protista (eukaryotic single-celled organisms with membrane-bound cell organelles. Amoeba and slime mold are examples of these); Fungi (multicellular, non-motile organisms that reproduce asexually and derive their nutrients from decaying organic materials); Plantae (multicellular, non-motile organisms that contain chlorophyll, the sunlight-catching pigment capable of synthesizing food by means of photosynthesis); and Animalia (multicellular, motile organisms that cannot synthesize food and whose mode of nutrition is ingestion).

In more recent years, scientists have further divided the Monera kingdom into Eubacteria and Archaebacteria. The former refers to true bacteria, the latter comprises bacteria-like organisms that inhabit, as previously mentioned, extreme environmental conditions such as alkaline lakes, sea ice, sulphuric springs and deep sea vents.

Viruses and other non-cellular entities are not generally considered "living" organisms, though at times the terms Acytota or Aphanobionta are used to refer to the viral domain.

[110] Although these three kingdoms are the ones that come most readily to mind whenever we think of living organisms, it is worthwhile to remember that they, as well as the protista, all evolved out of non-nucleated bacteria. These continue to dominate the planet in sheer number, variety and spread. They are, moreover, essential to the life processes of all other organisms.

[111] Long believed to be a uniquely-human characteristic, the use and even manufacture of tools is now known to occur in other species as well. Otters smashing mollusks with rocks and crows dropping pebbles in deep vessels to raise the water level are well-known examples of tool

use. Tool-*making* animals include the woodpecker finches of the Galapagos Islands, which use cactus spines to pry grubs out of branches, often shortening the spines for easier manageability, and carrying them from branch to branch as needed.

Wild chimps construct tools from twigs to "fish" ants and termites from holes, extract honey from beehives, dig up edible roots, and even as levers to open boxes of bananas left by scientists. The use of these tools requires the chimp to find a suitable twig, break it, strip away the leaves, and insert it into the nest while standing as far away as possible to avoid being bitten. This behaviour is not instinctual, and young chimps have been observed to play-mimic adults in such activities.

[112] This version of the hominid story is what we currently have, but it is a provisional account at best, for paleoanthropology is ever surprising. For example, in 2003, the partial remains of Homo floresiensis, a previously unknown species of dwarf hominids, was discovered on an island in the Indonesian archipelago.

Some of the extinct relatives of our species include Sahelanthropus tchadensis, Orrorin tugenensis, Ardipithecus ramidus, Australopithecus anamensis, Australopithecus afarensis, Kenyanthropus platyops, Australopithecus africanus, Australopithecus garhi, Australopithecus sediba, Australopithecus aethiopicus, Australopithecus robustus, Australopithecus boisei, Homo habilis, Homo floresiensis, Homo georgicus, Homo rudolfensis, Homo erectus, Neanderthal, Denisovan.

[113] A study published in *Science* asserts that language originated in central/southwest Africa and then spread across the globe through migration. The study compared phonemes (the smallest sounds which differentiate meaning, such as "sh" in "shook"; replace it with "l" or "b" and the result is another word) from 504 of the world's approximately 6500 living languages. It found that the more distant a language is from central/southwest Africa, the smaller the phoneme diversity. This implies that, just as the further one is from Africa, the poorer the genetic diversity due to inbreeding among relatively small recently-arrived migrant groups, so it is with phonemic diversity. ("Babel or babble?" *The Economist.* April 14, 2011)

[114] A recent study conducted on the Mundurucu tribe of the Amazon suggests that geometric intuition also is innate. (blogs.physicstoday.org/

---

newspicks/2011/05/geometry-skills-are-innate-acc.html)

[115] The ease with which young children acquire a language suggests for Chomsky that humans are born with an instinct for language. Specifically, he argues that the human mind has an innate "language acquisition device" that primes it for language and puts constraints on its range of possibility. He called the features that result from these constraints "universal grammar". It is UG that provides the framework upon which a particular language (eg. Swahili) hangs and without which no language is possible.

[116] Humankind could not survive without the innumerable resources and processes supplied by the ecosystem enveloping the planet. These include food, of course, but also the purification of air and water, waste decomposition, crop pollination, pest and disease control, nutrient dispersal and cycling, etc. These are but a few of the known examples of how we benefit from a healthy and diversified ecosystem. In fact, diversity is so vital that the 2010 UN report *The Economics of Ecosystems and Biodiversity* stated that saving species is of greater urgency than battling climate change. The ratio of costs to benefit in preserving biodiversity was cited in the report to be between 1:10 and 1:100. (www.guardian.co.uk/environment/2010/may/21/ biodiversity-un-report)

[117] The Jesuit philosopher-paleontologist Pierre Teilhard de Chardin (1881-1955) reached a rather different conclusion than Russell. Far from being a purposeless "outcome of accidental collocations of atoms," for de Chardin the universe and its evolution are tending ever towards a more perfect end. All creation is ascending towards Omega, and though chance and contingency are factors in evolution, its general direction is benevolently steered by the attractive force of the Omega Point.
  There are four stages in the evolution of the cosmos in de Chardin's metaphysical hypothesis: the prelife geosphere; the biosphere, which came to be with the evolution of life; the noosphere, with the emergence of human beings and rational thought whereby the earth "finds its soul," and man, with his reason, is able to direct evolution and reduce the force of chance; and the Cosmic Christ, the perfection of evolution. At this stage, the cosmos in all its particularities becomes united in one all-encompassing embrace, each particularity touching every other and

conscious of itself as a centre of consciousness amidst a cosmos of centres of consciousness.

Less mystically but in a similar vein, the Nobel biologist Christian de Duve imagines our species evolving into a planetary super-organism, which he calls the "human hive".

Futurist/economist Jeremy Rifkin suggests that we may be on the threshold of the Third Industrial Revolution whereby everyone is teleconnected and part of a renewable-energy power grid that spans the planet, thereby "extending the central nervous system of billions of human beings" into a kind of post-McLuhanesque globurbia.

[118] "The mind of God." *Omni*. February, 1992

[119] Although I am speaking metaphorically here, there is a scientific-philosophical notion referred to as the *Anthropic Principle* which posits essentially this very thing. Here follows a brief account of it.

Life is possible because the so-called universal constants of physics and the parameters for the universe and the Earth lie within certain highly restricted ranges. This is generally acknowledged.

> The origin and evolution of life has only taken place
> because the fundamental constants have produced the
> universal chemistry [that] ensured that a universe benign
> toward life has developed. (Explorations: 11)

If our universe is merely one of countless "bubble" or "parallel" universes in each of which different laws of physics operate and in which life may or may not have evolved, then it is only a fluke that life and mind evolved as it did in ours, and we are merely lucky to find ourselves in just such a universe. If, on the other hand, ours is necessarily the only or one of a very few universes in which life and mind could have evolved, then our universe was fit for life and mind from its very beginning. Why? Because the elements that play a key role in life processes are produced in stars and dispersed through space only at the end of the stars' life cycle, all according to physical laws involving very precise constants and billion-year long processes. Consequently, thinking backwards, the existence of life and the stars and galaxies that gave rise to it, are the result of a trajectory that traces back to the Big Bang, if not earlier (see endnote 99). This is how scientists (including Fred Hoyle, Frank Tipler, John Barrow, and John Wheeler) have come to explain the exquisitely fine-tuned "coincidences" that are necessary for life. "A life-giving factor lies at the centre of the whole machinery and design of the world." (Wheeler, as quoted in Ross)

The Anthropic Principle essentially says that life is so improbable that its existence can "explain" the universe by placing constraints on the parameters of the physical laws that describe it and make life possible. Over the years different versions of the principle have been proposed, from the *weak* (which states that the universe must be such as to allow for the existence of intelligent life, for here we are!) to the *strong* (that the universe must have properties that make intelligent life inevitable) to the *final* (that once intelligent life has come into existence in the universe, it will never disappear).

The Anthropic Principle germinated in 1961 when the physicist Robert H. Dicke used it to explain the coincidences among certain ratios that fix the relative strengths of the fundamental constants of nature. For example, the ratio of the electrical to the gravitational forces between a proton and an electron ($\approx 10^{40}$) is also the ratio of the age of the universe to the time it takes light to cross an atom, and also equal to the square root of the number of protons in the universe at the present time. Had this ratio been slightly different, life could not have arisen, he realized.

Since Dicke's time, over a hundred other fundamental constants have been identified. These include:

1) The fine-structure constant $\alpha$, (a strange brew of the speed of light, Planck's constant, the charge of the electron, and the number $\pi$) which determines the strength of the electromagnetic attraction between electrically charged particles. If this number were 4% smaller or bigger than it is, stars would not be able to sustain the reactions that produce carbon and oxygen atoms and, hence, the chemistry required for life (at least as we know it) would not be possible.

2) The strength of the so-called gravitational coupling constant, which determines the gravitational attraction between charged elementary particles. If this number had been slightly stronger, star formation would have proceeded more efficiently and all stars would have been at least 140% more massive than the sun. Such stars burn too rapidly and unevenly to maintain life-supporting conditions on orbiting planets. If, on the other hand, the constant had been slightly weaker, all stars would be 20% smaller than the sun. Such stars burn long and evenly, but no heavy elements essential for building planets or sustaining life could be created within them and then disseminated via supernova explosions.

3) The strong interaction or strong force, which characterizes the binding energies of atomic nuclei. It is defined in terms equivalent to the mass of the hydrogen atom that is released as energy (according to Einstein's $E = mc^2$ formula) when two hydrogen atoms fuse to form a helium atom. If its value were 0.1% smaller, the universe would contain nothing but hydrogen. If it were 0.1% bigger, all the hydrogen would have fused into heavier elements. In either case, the chemistry necessary for life would not have come about.

4) The ratio of electron to proton masses determines the characteristics of the orbits of electrons around the atom's nucleus. A proton is 1836 heavier than an electron. If it were slightly more massive, atoms could not share electron orbits with other atoms. If any less massive, no electrons could be held in orbit around the nucleus. In either case molecules could not form and life would be impossible.

5) The density parameter $\Omega$ determines the geometry of the universe and, therefore, the rate of expansion of the universe. Had the number been a fraction smaller, the universe would have expanded too quickly for gravity to draw hydrogen atoms together and thereby form stars; if a fraction larger, the universe would have collapsed back onto itself before any stars could have settled into a stable burning phase. In either case, no planetary systems capable of supporting life would have resulted. (Ross: 1-7)

It is interesting to note that recent data suggests that the fine structure constant $\alpha$, a combination of the electron charge, the speed of light and Planck's number, may not be constant after all. Scientists have evidence that its value changes ever so slightly (0.0006% to be precise) depending in which direction of the universe one is looking. This has profound implications, for it would imply, again, not only that the universe stretches far beyond anything previously imagined ("stretch[ing] not just millions of times further than our currently observable domain, *but millions of powers of ten* further" – Rees, 13), but that the laws of physics can vary within it. Instead of the entire universe being fine-tuned for life, humanity happens to find itself in a bubble or patchwork of space-time where the fundamental constants happen to be just right for life to evolve. ("Ye Cannae Change the Laws of Physics — Or Can You?" *The Economist*. August 31, 2010)

Rather than constants varying over space, Rupert Sheldrake, following Nobel Physicist Paul Dirac, suggests that they evolve over time. The reason for this is that many of the constants are ratios that involve the universe's age. As the latter changes, so necessarily must the ratios and, hence, the value of the constants. Evolving constants would solve the problem of how to account for the laws of nature before the creation of the universe, for they would no longer have to have existed in some pre-Big Bang platonic realm. Rather, they were created with the universe and developed along with it.

Regardless whether the universal constants are fixed or not, the improbable fact of finely tuned ratios is very difficult to explain without some assumption of design, other than to posit an infinite number of multiverses, all governed by infinite combinations of infinitely-varied constants(!); or that our universe and its laws are only local phenomena of a much vaster megaverse; or, as some are now proposing, that the universe may be governed by a feedback loop that operates forward and backward in time and thereby creating its own efficacious conditions.

However one ponders the matter, it is an astonishing mystery.

[120] Contrary to the insistence of the biotechnology industry, there is now incontrovertible proof that genes *can* jump from species to species via bacterial vectors. In fact, according to biologist Stephen L. Coles, one of the surprising lessons of the Human Genome Project is that bacterial gene cartridges of as many as "200 genes in one fell swoop" can be transferred as a block into human chromosomes. (Coles: 3)

The danger is that once such a transfer of genetic material has occurred somewhere in the biosphere, it could end up anywhere in the biosphere with unforeseeable and, perhaps, dire irreversible consequences.

[121] $2^{336} \approx 10^{101}$ whereas there are an estimated $10^{80}$ protons in the visible universe (Sagan: 219). Population growth is generally modelled by exponential functions, a type of non-linear function. In the example given, the number of bacteria after one week is $N = 2^{10,080/30}$ where N is the resulting number of bacteria, 2 is the base of the exponential growth (bacteria are doubling), 10,080 is the number of minutes in one week, and 30 is the doubling period (the number doubles every 30 minutes).

[122] Tat Tvam Asi (तत् त्वम् असि or तत्त्वमसि), is a Sanskrit sentence that occurs in the Hindu scripture *Chandogya Upanishad*. The meaning of the sentence is that the Atman, i.e. the Self in its essential, purest sense, immanent in every being, is identical with Brahman, the Ultimate Reality that is the ground and origin of all that was, is, and ever will be.

## Bibliography

Adler, Mortimer J. and Van Doren, Charles, ed. *Great Treasury of Western Thought*. (New York: R. R. Bowker Company, 1977)

Andrewartha, H. G. and Birch, L. C. *The Ecological Web*. (Chicago: University of Chicago Press, 1984)

Attenborough, David. *Life on Earth*. (Toronto: Little, Brown & Co., 1979)

Attenborough, David. *The Living Planet*. (Toronto: William Collins Sons & Co., 1984)

Augros, Robert M. and Stanciu, George N. *The New Story of Science*. (Toronto: Bantam Books, 1984)

Barrow, John D. *The Constants of Nature*. (New York: Pantheon Books, 2002)

Bateson, Gregory. *Mind and Nature: A Necessary Unity*. (New York: E. P. Dutton, 1979)

Berry, Thomas. *The Dream of the Earth*. (San Francisco: Sierra Club Books, 1988)

Berry, Thomas and Swimme, Brian. *The Universe Story*. (New York: HarperCollins, 1992)

Brown, Lester. *Eco-Economy: Building an Economy for the Earth*. (New York: W.W. Norton, 2001)

Calder, Nigel. *Timescale*. (London: The Hogarth Press, 1984)

Capra, Fritjof. *The Turning Point*. (New York: Simon and Schuster, 1982)

Capra, Fritjof. *The Web of Life*. (New York: Doubleday, 1996)

Capra, Fritjof and Steindl-Rast, David. *Belonging to the Universe*. (San Francisco: HarperCollins, 1991)

Carson, Rachel. *Silent Spring*. (Boston: Houghton Mifflin, 1962)

Chomsky, Noam. *Language and Problems of Knowledge*. (Cambridge, Mass.: MIT Press, 1988)

Coles, Stephen L. *A Review of the AAAS Seminar "Beyond the Human Genome."* (www.grg.org)

Davies, Paul. *The Cosmic Blueprint*. (London: Penguin Books, 1987) de Duve, Christian. *Vital Dust: Life As A Cosmic Imperative*. (New York: Basic Books, 1995)

De Quincey, Christian. *Review of "The Reflexive Universe"* (www.deepspirit.com)

Deutsch, David. *The Fabric of Reality*. (London: Penguin Books, 1997)

Devlin, Keith. *The Math Gene*. (Great Britain: Basic Books, 2000)

Diamond, Jared M. *Collapse: How Societies Choose to Fail or Succeed*. (New York: Viking, 2005)

Dykes, Nicholas. *A Critical Assessment of the Philosophy of Karl Popper*. (www. capital.demon.com)

*Explorations: Currents in the Interface of Science and Religion*. (John Templeton Foundation: 2000, V5, #2)

Ferris, Timothy, ed. *The World Treasury of Physics, Astronomy, and Mathematics*. (Toronto: Little, Brown and Co., 1991)

Gore, Albert. *An Inconvenient Truth: The Planetary Emergency of Global Warming and What We Can Do About It*. (Emmaus, Pa.: Rodale Press, 2006)

Goswani, Amit. *The Self-Aware Universe*. (New York: Penguin Putnam, 1993)

Gould, Stephen Jay. *The Book of Life*. (New York: WW Norton, 2001)

Gould, Stephen Jay. *Evolution: The Triumph of an Idea*. (New York: Harper Collins, 2001)

Greene Brian. *The Elegant Universe*. (New York: WW Norton & Co., 1999)

Greene Brian. *The Hidden Reality: Parallel Universes*. (New York: Alfred A. Knopf, 2011)

Grof, Stanislav. *The Holotropic Mind*. (San Francisco: HarperCollins, 1993)

Greenstein, George. *The Symbiotic Universe*. (New York: William Morrow & Co., 1988)

Haberman, Arthur. *The Making of the Modern Age*. (Toronto: Macmillan, 1977)

Harman, Willis W. and Sahtouris, Elisabet. *Biology Revisioned*. (Berkeley: North Atlantic Books, 1998)

Hawking, Stephen. *A Brief History of Time*. (Toronto: Bantam Books, 1988)

Holland, John H. *Emergence*. (Reading: Helix Books, 1998)

Horgan, John. *The End of Science*. (New York: Addison-Wesley Publishing, 1996)

Hutchins, Robert M. and Adler, Mortimer J. "What is Imagination?" *Gateways to the Great Books*, Vol I. (Chicago: Encyclopedia Britannica, Inc., 1963)

Kaku, Michio. *Hyperspace*. (Oxford: Oxford University Press, 1994)

Lanza, Robert P. *Biocentrism: How Life and Cosciousness Are Keys to Understanding the True Nature of the Universe*. (Dallas: BenBella Books, 2009)

Lomborg, Bjørn. *The Skeptical Environmentalist*. (New York : Cambridge University Press, 2001)

Margenau, Henry and Varghese, Roy Abraham, ed. *Cosmos, Bios, Theos*. (La Salle, Illiois: Open Court, 1992)

McBryde, W. A. E. and Graham, R. P. *The Outlines of Chemistry*. (Toronto: Clarke, Irwin & Co. Ltd., 1966)

Murchie, Guy. *The Seven Mysteries of Life*. (Boston: Houghton Mifflin Co., 1978)

Oreskes, Naomi and Conway, Erik M. *Merchants of Doubt*. (New York: Bloomsbury Press, 2010)

Paster, Robert. *New Physics and the Mind*. BookSurge, 2006

Prigogine, Ilya. "Interview." *Omni Magazine*. (www.omnimag.com: May, 1983)

Rees, Martin. *Just Six Numbers*. (Great Britain: Basic Books, 2000)

Rifkin, Jeremy. *The Empathic Civilization*. (New York: J. P. Tarcher/Penguin, 2009)

Ross, Hugh. *Design and the Anthropic Principle*. (www.reasons.org)

Sagan, Carl. *Cosmos*. (New York: Random House, 1980)

Sheldrake, Rupert. *A New Science of Life*. (Los Angeles: J. P. Tarcer, 1981)

Sheldrake, Rupert. *The Rebirth of Nature*. (Toronto: Bantam Books, 1992)

Sinetar, Marsha. *Developing a 21$^{st}$ Century Mind.* (New York: Villard Books, 1991)

Singer, Peter. *Animal Liberation.* (New York: Random House, 1990)

Singer, Peter. *One World: The Ethics of Globalization.* (New Haven: Yale University Press, 2004)

Smolin, Lee. *The Life of the Cosmos.* (New York: Oxford University Press, 1997)

Sperry, Roger. "Interview." (*Omni Magazine*: August, 1983)

Sperry, Roger. *Science and Moral Priority.* (New York: Prager Scientific, 1985)

Strong, Maurice. *Where On Earth Are We Going?* (Toronto: Knopf, 2000)

Swimme, Brian. "Interview." *EarthLight Magazine.* (www. earthlight.org: Summer, 1997)

Talbot, Michael. *The Holographic Universe.* (New York: HarperCollins, 1991)

Thomas, Lewis. *The Lives of a Cell.* (New York: Bantam Books, 1975)

Vygotsky, Lev. *Thought and Language.* (Cambridge: The MIT Press, 1999)

Weyer Jr., Edward M., ed. *Strangest Creatures on Earth.* (New York: Sheridan House, 1953)

Watson, James D. *The Double Helix.* (New York: Simon & Schuster, 2001)

Weinberg, Steven. *Dreams of a Final Theory.* (New York: Vintage Books, 1992)

Weisman, Alan. *The World Without Us.* (Toronto: HarperCollins, 2007)

*Wikipedia* (en.wikipedia.org)

Wilber, Ken, ed. *The Holographic Paradigm and Other Paradoxes.* (Boulder: Shambala, 1982)

Wilber, Ken. *A Brief History of Everything.* (Boston: Shambala, 1996)

Wilson, Edward O. *Consilience.* (New York: Vintage Books, 1999)

## *Index*

α, 196, 197
Adler, Mortimer J., 116
Aesop, 118
amino acids, 55
amphibians, 132
anaerobic bacteria, 128
animals, 130, 131
Anthropic Principle, 195, 196
ants and termites, 154
Arnold, Matthew, 115
Arctic Monitoring and Assessment Program, 165
Aristotle, 35, 36, 171, 178
Aspect, Alain, 27, 180
ATP, 127, 128, 129, 131
*Australopithecus africanus*, 144
autopoiesis, 56, 58, 60, 70
Babbit, Bruce, 166
Bacon, Francis, 10, 168
bacteria growth 73, 155
Bakker, Robert, 134
Barfield, Owen, 96
Bateson, Gregory, 61, 173
beetles and flies, 154
Behaviourism, 21, 85
Bell's Theorem, 10, 28
Benard, Henri, 52
Benard Instability, 53
Bergson, Henri, 86, 180

Berkeley, George, 179
Berman, Morris, 171
Berry, Thomas, 106, 145, 150, 1841, 186
bicameral brain, 180
Big Bang, 123, 124, 187, 189, 198
Biocentrism, 78
biosphere, 2, 14, 67, 69, 71, 72, 100, 101, 113, 127, 147, 152, 154, 191, 198
Bohm, David, 29, 170
Bohr, Neils, 170
bonobo, 157
branes, 188
Broad, C.D., 32
Broca's Area, 143
Brown, Lester, 15
Bruno, Giordano, 178
Buber, Martin, 150
Bush, George, 105
Cairns, John, 74
Campell, Joseph, 123, 144
Cancún summit, 182
Capra, Fritjof, 45, 106
Carson, Rachel, 15
Cartesian-Newtonian, 9, 17, 28, 31, 62, 85, 101, 110, 112, 116, 169, 185
Chaos Theory, 10, 173
Chicxulub, 133

Chomsky, Noam, 144
Chopra, Deepak, 81, 1819
cicadas, 44
Club of Rome, 16
cockroaches, 72
Coles, Stephen L., 198
Collective Unconscious, 172
Complementarity, 170
consciousness, 19, 21, 25
constant, 169, 196
Copernicus, 111
cosmic religious feeling, 148
Crick and Watson, 40
Crick, Francis, 175
*Critique of Pure Reason*, 169
cybernetics, 9, 47, 49, 58, 61, 174
Dalton, John, 18
Darwin, Charles, 71, 174
Davies, Paul, 149
Dawkins, Richard, 40, 172
de Chardin, Pierre Teilhard, 194
de Duve, Christian, 195
deniers, 166
Devlin, Keith, 143
Dicke, Robert H., 196
Dirac, Paul, 197

dissipative structures, 50, 51, 53, 60, 191
DNA, 40, 41, 129, 178
Driesch, Hans, 36
Earthrise, 15
Easter Islands, 167
Eccles, John, 25, 94, 95
ecological footprint, 11
Economics of Ecosystems and Biodiversity Report, 3
Ecosophical Paradigm, 5, 108, 111, 112, 114, 118, 149, 151, 160, 184
ecosystem, 14, 31, 33, 34, 47, 156, 162, 194
Eddington, Arthur, 85
Ehrlich, Paul, 13
Eigen, Manfred, 54
Einstein, Albert, 17, 18, 20, 28, 165, 189
Eliade, Mircea, 119
Eldredge, Niles, 174
emergent properties, 32, 79, 90, 176
Enlightenment, 21, 25
entelechy, 37, 171
equilibrium, 49, 51, 52, 53, 54, 60, 125
*Escherichia coli*, 175
eukaryotes, 130
Existentialism, 100

exoplanets, 190
feedback loop, 42, 48, 51, 53, 58, 67, 154, 173
Feynman, Richard, 186
Fleischaker, Gail, 72
fungi, 131, 160
Gage, Phineas, 82
Gaia Hypothesis, 67, 68, 69, 70, 71
Gödel's theorems, 9, 169
Gould, Stephen Jay, 142, 174
gravitational waves, 171
Great Chain of Being, 171
green political parties, 16
Greenstein, George, 82
Grof, Stanislav, 92, 181
Haldane, J. B. S., 148
Hall, Barry, 74
hard problem, 75
Harvard U. study, 172
Hazlitt, William, 172
Heisenberg, Werner, 20, 169
Heisenberg's Principle of Uncertainty, 169, 179
Henderson, Lawrence, 46
Higgs boson, 188
holarchy, 31, 33, 72, 74, 75, 97, 131
Holism, 185
hologram, 29, 170
holon, 31, 32, 33, 34, 63, 71, 76, 77, 88, 126

homeostasis, 49
*Homo erectus*, 137 193
*Homo habilis*, 137, 193
*Homo sapiens*, 137
*Homo sapiens sapiens*, 138
Hume, David, 178
hurricanes, 161, 163
Hurricane Katrina, 157
Husserl, Edmund, 180
Huxley, Aldous, 93
hypercycles, 50, 54, 56, 60, 125, 191
Hypothesis of Formative Causation, 42
Information Theory, 9, 170
International Union for Conservation of Nature (IUCN), 3
Jean, James, 77
Jung, Carl, 119, 172
Kaku, Michio, 188
Kant, Immanuel, 170
Kingdoms of Life, 132, 192
Koestler, Arthur, 31
Korzybski, Alfred, 67
Kosmos Story, 144, 149, 153, 156, 186
Kuhn, Thomas, 110, 185
languages, 193
Lanza, Robert, 78
Lashley, Karl, 181

*Limits to Growth, The,* 16
Lindstrom, Martin, 183
Linnaeus, Carl, 192
Lomborg, Bjørn, 167
Loorz, Alec, 186
Lovelock, James, 69
LSD-25, 86
Lundeen, Jeff, 179
Maturana, Humberto, 50, 56, 60, 62, 63
Margulis, Lynn, 66, 177
Maya, 167
McKenna, Terence, 192
Mead, Margaret, 169
meaning, 6, 9, 14, 97, 119, 122, 149, 151, 152, 155, 183
Mechanism, 35, 37, 38, 41, 45, 186
mechanomorphism, 167
megaverse, 187, 198
meme, 172
metapattern, 173
Millennium Ecosystem Assessment, 3
Miller, Stanley L, 55
Miller-Urey experiment, 55
mitochondria, 129, 177
Moche, 167
monarch butterflies, 44
Morgan, C. Lloyd, 35
morphic fields, 42, 43, 44

morphic resonance, 42, 43, 44, 173
M-Theory, 183, 188
Multiple Personality Disorders, 180
Mundurucu, 193
Naess, Arne, 69, 150
NASA, 66
Navajo, 179, 183
new ethic, 1, 2, 5, 112
Newton, Isaac, 21
Nobel laureates, 1
Omega Point, 194
panentheism, 179
panpsychism, 75
panspermia, 175
pantheism, 179
paradigm, 110, 112
Penfield, Wilder, 92, 181
perennial philosophy, 95
phonemes, 193
photosynthesis, 127, 192
Planck, Max 18, 77, 169, 185, 189, 196
plants, 25, 67, 130, 131, 133, 160, 192
Popper, Karl, 21, 96, 181
Pribram, Karl, 170
Prigogine, Ilya, 51
primates, 31, 134, 135, 154, 172
prokaryotes, 66, 72, 73, 126, 129, 155, 177, 192

prosimians, 135
Protagoras of Abdera, 168
*qualia*, 180
Quantum Field Theory, 188
Quantum Mechanics, 9, 46, 74, 80, 81, 82, 85, 169, 178, 179, 187
quarks, 32, 45, 59, 124, 188
rats, 168
recursive function, 173
Relativity Theory, 10, 17, 28, 85
reptiles, 132, 133, 164
respiration, 132, 133
ribosomes, 130, 177
Rifkin, Jeremy, 195
Rio Earth Summit, 103
RNA, 56, 178
Rorty, Richard, 120
Russell, Mike, 56
Sahtouris, Elisabet, 77
Santiago Theory of Cognition, 61, 62, 65
Schmidt, Eric, 186
Schröedinger's equation, 83, 84
scientific revolution, 2
Second Law of Thermodynamics, 80
Sheldrake, Rupert, 42, 168

*Silent Spring*, 15
Sinetar, Marsha, 185
Sixth Great Extinction, 4, 163
Smith, Houston, 171
space-time, 28, 79, 93, 123, 186, 197
Sperry, Roger, 24, 25, 27, 86, 87, 89, 91, 93, 95, 97, 100, 102, 104, 113
Spinoza, 179
split-brain phenomenon, 24
St. Vincent Millay, Edna, v
Standard Model, 188
Stern, Lord Nicholas, 183
Steinberg, Aephraim, 179
Steindl-Rast, David, 184
Stowe, Harriet Beecher, 120
Strong, Maurice, 12
subatomic particles, 25, 189
sun, 190
supernovae, 190
Swimme, Brian, 151
symbiogenesis, 177
symbiosis, 177
Systems Theory of Life, 45, 57, 58
Tansley, Arthur G., 47
termites, 32, 33, 51

The Law of Sensitive Dependence on Initial Conditions, 174

*The Selfish Gene*, 172

*The Structure and Significance of the Consciousness Revolution in Science*, 24

*The Structure of Scientific Revolutions*, 108

Theory of Everything (TOE), 188

Theory of Punctuated Equilibria, 174

therapsids, 133

Thoms, Michael, 9, 106

Thomas, Lewis, 71

Thoreau, Henry, 150

three worlds of reality, 181

Traditional Mind, 185

21$^{st}$ - Century Mind, 185

universal grammar (UG), 194

University of Toronto, 179

upward causation, 25

Urey, Harold C., 55

Varela, Francisco, 50, 56, 57, 58, 60, 62

Vernadsky, Vladimir. I., 177

Vico, Giambattista, 22

viruses, 62, 73, 178, 192

Vitalism, 35, 41, 45, 80

von Neumann, John, 61, 174

Vygotsky, Lev, 176

Wald, George, 85

water, 1, 25, 32, 55, 69, 128, 131, 190

Weber, Max, 104

Wells, H. G., v

Wheeler, John, 77, 81, 195

Whitehead, Alfred North, 85

Wiener, Norbert, 48, 49, 61, 174

Wigner, Eugene, 84

Wilber, Ken, 121, 171, 182

Wilson, E. O. 118, 150

Wolfram, Stephen, 176

*World Scientists' Warning to Humanity*, 1, 163

Worldwatch Institute, 15

Yau, Shing-Tung, 187

Yucatán, 133

Zhabotinsky Reactions, 53

$\Omega$, 197